Man's Mathematical Models

Fundamental Concepts for the Nonmathematician

Man's Mathematical Models

Bill Williams, Ed. D., &
Gwen Crotts, M.A.

Nelson-Hall Chicago

ISBN: 0-88229-110-6

Library of Congress Catalog Card Number: 73-93104

Manufactured in the United States of America

Library of Congress Cataloging in Publication Data

Williams, Bill, 1940–
Man's mathematical models.

Includes bibliographical references.
1. Mathematics—1961– I. Crotts, Gwen, 1946–joint author. II. Title.
QA39.2.W558 512 73-93104
ISBN 0-88229-110-6

Contents

Preface

There was a period in educational history when college students who were majoring in liberal arts were generally required to take the same type of introductory mathematics course that was required of students majoring in mathematics or science. To a large degree, the liberal arts student rebelled by turning off to mathematical activity and thus forfeiting the power and beauty in the habits of thought of mathematics. In recent years, the mathematical community has awakened to the realization that it is our responsibility to provide mathematical experiences that will not only challenge and interest the vast number of nonscience students but which will also make mathematics more humanistic in nature.

Many fine materials designed specifically for the liberal arts student have appeared in recent years. The authors feel that the present work will provide still another dimension of mathematical opportunity for the liberal arts student by revealing components of mathematical endeavor that can serve a nonscience student extremely well regardless of his specialized interest. We believe these components are found in the methods and processes of mathematics.

To enable the liberal arts student to operate well in his own field of interest, he should gain the advantage of a field trip into the mathematical arena to gather tools in which the mathematician specializes. Some important tools that serve to broaden the educational base of all students and give them greater insight into other areas of learning are: the conceptualization of structure as it relates to a body of thought, the cultivation of the powers of abstract reasoning, and an understanding of the deductive process.

Students come to college with tremendously varied backgrounds in mathematics. Keeping this in mind throughout this book, the authors have tried to make only the most elementary assumptions regarding the student's background. We are well aware that many materials written for the liberal arts student are much too ambitious, and we have kept a priority of careful selection of topics and ease of development throughout the years of preparing the notes, which were the origins of this manuscript. A few topics, which are traditionally included in particular sections, are omitted for this reason.

Typically, a mathematics text for liberal arts students covers many topics including sets, logic, number systems, abstract algebra, introduction to computer programming, limits, intuitive calculus, probability, and statistics. The authors' experience has been that the average student cannot possibly absorb or relate so diverse a presentation of mathematics. Hence, we have narrowed the selection, and we relate the selected topics to each other with the emphasis on the unifying threads that run throughout and carry over into other areas of endeavor. Our motivation is not centered upon a historical presentation nor in the real world application of mathematics. We do use these features when appropriate, but we have found that the elegance of the reasoning process is motivation enough in learning mathematics provided the level of difficulty is not too great and the student can participate in the process.

Because of the nature of the material presented, it is beneficial to make many avenues of learning open to the students. We have included the following:

1. Use of a unifying theme, that is, set theory concepts
2. Use of many examples to convey the concepts, some familiar to the students and some new

3. Use of diagrams as visual representations of concepts
4. Use of detailed verbal descriptions of the concepts
5. Use of symbols to formulate the concepts

The authors feel that in addition to making many avenues of learning available, there is no substitute for enthusiasm in the presentation of the material to help a student gain an appreciation for mathematics. The style of writing in these materials is intended to pass along this enthusiasm, and every opportunity is taken to make the materials readable without compromising appropriate content.

In recent years, most students have been exposed to the basic terminology and concepts of sets. Many students have a good understanding of these concepts, while others do not. Hence, we have written Chapter 1 on sets in such a manner as to provide for differences of mathematical understanding regarding sets. The labeling of the sections of Chapter 1 is a means of acknowledging that the student has been exposed to some of the materials before, and we have found that our students greatly enjoy the titles of these sections. If a class needs the very basic notions of sets, the first six sections provide for this need. If a class has a good background in sets, the first six sections can be read lightly or the work can begin with Section 7. If a class has an excellent background in set concepts, the work can start with Chapter 2.

The authors feel that Chapters 2, 3, 4, and Sections 1 and 2 of Chapter 5 provide the basic thrust of these materials. Chapter 1 can be included according to the above criterion, while Sections 3 through 7 of Chapter 5 and Chapter 6 will provide for a broader view of mathematics and number systems if time permits the use of them. The authors have found that a class with a superior background in mathematics can cover most of the materials in this text in one semester, while a less well prepared group may cover only through Section 2 of Chapter 5. In any case, by covering Chapters 2, 3, 4, and Sections 1 and 2 of Chapter 5, the major emphasis of these materials will be realized.

Of all the topics covered in a class of this nature, it has been the authors' experience that the greatest level of interest expressed on the part of the students has been in the area of symbolic logic. The material is new to them, does not have the typical computational content of past mathematical experiences and has implications for generalization to their particular area of interest. Usually, the basic principles of symbolic logic are presented as only a set of rules, defined with little motivation and tied in minimally with sets or with other mathematical ideas. The authors have found through experience that little is gained by students from this manner of presentation. Hence, in Chapters 2 and 3, we have made the area of symbolic logic and deductive reasoning a major development of this book. Since interest is so high in the area of symbolic logic, it is only good pedagogy to carry it to a natural conclusion in teaching deductive reasoning. We do not feel that symbolic logic is in and of itself "all important," but we do believe that it is a vehicle to help students conceptualize the deductive method.

Chapter 4 presents the notion of function in some detail. This material is developed slowly and painstakingly so that the student can gain maximum facility with the concept. We treat together the basic ideas of functions and binary operations, thus letting the student center on these concepts before exposing him to their applications to groups and other abstract structures. Chapter 4 is somewhat of a change of pace from Chapters 1, 2, and 3, but the underlying notion of sets is carried through in a straightforward manner.

Chapter 5 presents the student with the opportunity to put to use all of the previously covered material. The deductive process is emphasized in the context of groups, and set notions and function concepts are tied in naturally with the group as a deductive mathematical system. We have found that the student is not overwhelmed by the abstract nature of this system, because he has been carefully prepared for the undertaking in preceding chapters. Usually, students are exposed to the deductive process as it relates to groups without sufficient preparation. That is, they must struggle with the deductive reasoning process concept, the set concept, and the function and binary operation concepts, all at the same time without adequate familiarity with each concept before bringing them all to bear on this particular system. We have experienced success in a slower, more painstaking presentation, and we find that students enjoy the familiar arithmetic operations of addition and multiplication from the perspective of this system.

Throughout our work we use only the familiar number systems as examples and do not develop them rigorously. We find that rigorous development of something the students feel that they know intuitively can quickly turn off the enthusiasm that we plan to build for mathematics. But, by using algebraic systems with many models of representations including the number systems, the student can gain facility with the concepts of the number systems as well as facility with the deductive process.

The last chapter, Chapter 6, on rings and fields carries the materials previously presented to their natural conclusion as they relate to the deductive process, and representations are presented from many areas as well as from the familiar number systems. Some examples of near-rings are given to help the student see the importance of particular axioms.

To challenge students of varied ranges of ability, we include exercises of various levels of difficulty. The exercises occurring in this manuscript were selected by the authors on the basis of feedback obtained from students who used these materials in earlier manuscript form. The exercises are a tool to help the student sharpen his understanding of material developed and to expand upon material in the body of the text. Answers to selected exercises are included at the end of the book.

To the mature mathematical eye, various statements made regarding the concept of sets, the development of notation, and the accumulative nature of the materials presented may appear to be needlessly repetitious. The authors have found that keeping the student constantly aware of the interrelatedness of various ideas helps him to conceptualize the continuity of development. However, unless specific emphasis is placed on such relationships the novice in mathematics may not see them.

Suggested readings are given at the end of each chapter.

These readings either expand on topics presented in the chapter or relate some of the materials presented in the chapter to other areas.

The authors have found that the materials in this book may also be useful for prospective teachers of mathematics in both elementary and secondary schools. Many prospective teachers have told us that they have found the materials very helpful.

The authors wish to express their gratitude to all those who had a part in the preparation of this manuscript. In particular we wish to express our thanks to the students and colleagues who used these materials in various stages of development and for the valuable criticisms they offered. To Mrs. Barbara Hintze for her careful reading of the final manuscript and the galley proofs, we express our appreciation. Finally, we wish to say thank you to Mrs. Lorraine Dakin and Mrs. Leslie Dodrill for their conscientious typing of the manuscripts.

CHAPTER ONE

Sets

Section 1
Collector's Guide

Are you reading this book because you find mathematics exciting? Because the cover is so attractive you could not resist further investigation? Or because it is a class assignment (perish the thought)? In deciding to read this text, you may have gone through a process of thought similar to the following:

I could be participating in many activities. They are: (a) reading this material, (b) attending a social event, (c) reading other academic materials, (d) swimming, (e) eating, (f) working, (g) watching TV, (h) sleeping, or (i) playing tiddlywinks. At present, some of these activities would be more worthwhile than others: working, reading this material, and reading other academic materials. Having narrowed it down this far, I choose to read this material.

To help us keep track of the classifications of these activities, we can construct a diagram as shown in Figure 1.1.

Now let's suppose Jeff wants to make a date for a particular evening. What process does he go through? First, he considers all of the girls he knows. Then he selects the ones he would consider asking for a date. Having done this, he contacts a particular girl. We show this diagrammatically in Figure 1.2.

Notice that each of the above examples has the same pattern. First, each starts with a general collection of enti-

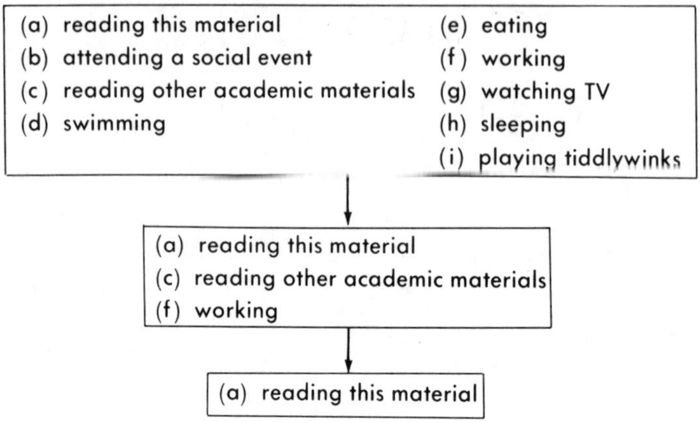

Figure 1.1

ties. From this collection, another collection is formed according to some criterion for classification. The second collection formed consists of a portion of the original general collection.

The words *classify* and *collection* are of central importance in mathematics. When we classify entities, we decide whether they are in a particular collection or not. The phrase *collection of objects* describes what mathematicians call a *set*. In *Roget's Thesaurus*, the words *collection* and *set* are given as synonyms; but, since describing a set in terms of a synonym does not extend our notion about the concept of set,

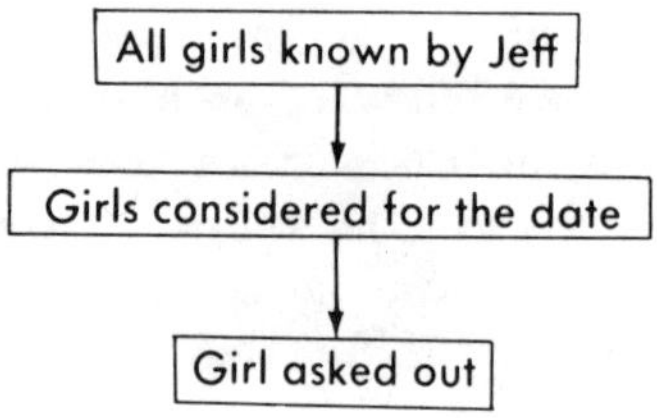

Figure 1.2

we gain only an intuitive feeling for the meaning of set. Mathematicians, however, concern themselves only with sets that are *well defined;* that is, sets that are so well described that we can tell whether or not a given entity is a member of a given set.

Let's look at some examples. The set that consists of all interesting courses offered in your school is not well defined since people's opinions as to what is interesting vary according to their backgrounds and their interests. The set that consists of the names of all states in the United States is well defined, since given any entity, we can determine whether or not it is a member of the set. Some additional examples of well-defined sets are:

1. All planets in the solar system.
2. All people in New York City.
3. All primary colors.

Now, let's go back and take a look at the pattern followed by our first two examples.

1. They started with a general collection of entities.

A collection or set that contains all of the objects under consideration for a particular discourse is called a *universal set.* We denote this set symbolically by U. Thus, in one case, our universal set consisted of the following activities: reading this material, attending a social event, reading other academic materials, swimming, eating, working, watching TV, sleeping, and playing tiddlywinks. In the other case, our universal set consisted of all of the girls Jeff knew. As observed, each illustration has its own universal set.

Do not think of a universal set as a set consisting of everything. This leads to logical paradoxes in set theory. (Your teacher can provide you with a discussion of this interesting topic, and additional sources are listed at the end of the chapter.)

2. From the general collection (universal set), another collection was formed according to some criterion for classification.

We note that a criterion for classification separates the universal set into two distinct groups: (1) a collection consisting of those items that satisfy the criterion and (2) a collec-

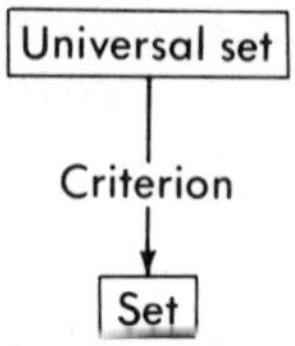

Figure 1.3

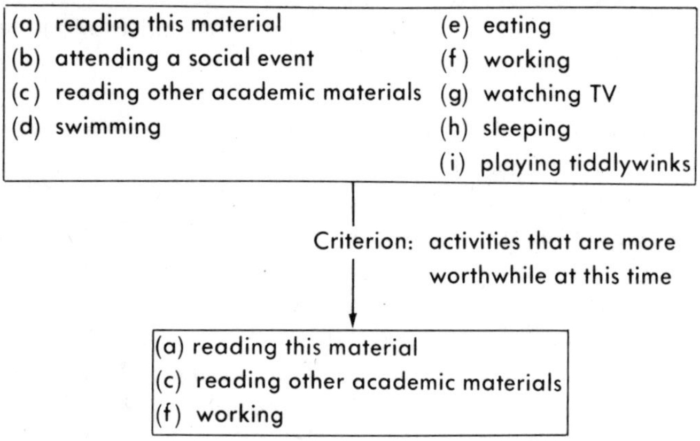

Figure 1.4

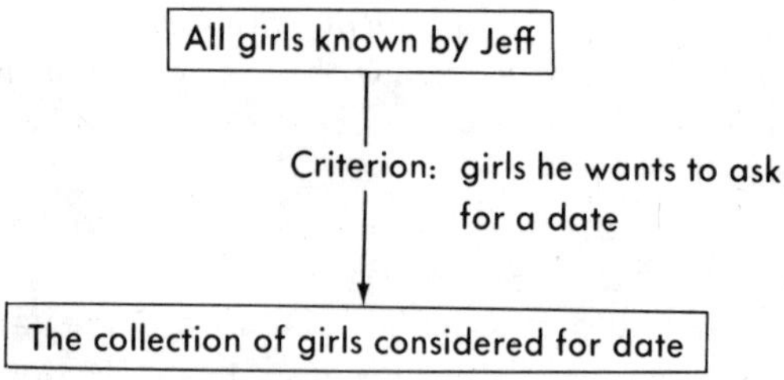

Figure 1.5

tion consisting of those items that do not satisfy the criterion. The collection of items that satisfy the criterion is the set of immediate interest. Figure 1.3 illustrates the pattern.

For example, using the first illustration, we can construct Figure 1.4.

Similarly, Figure 1.5 shows the pattern for Jeff's making a date.

In summary, we note that for each set under consideration there is an associated universal set and a well-specified criterion that, when applied to the universal set, separates it into two parts: the entities in the set under consideration and the entities not in the set under consideration. In general, the term *set* refers to a portion of a universal set, but it can refer to the universal set itself.

The individual members of a set are called *elements*. The symbol $\in$ is the symbol used to mean "is an element of," and $\notin$ means "is not an element of." Capital letters are used to designate sets, and small letters are used for the members of a set. Let A be a set. Then, if a is an element of A, we write "$a \in A$," and if a is not an element of A, we write "$a \notin A$."

Example 1: Suppose set A has as its elements the numerals 1, 2, and 3. Then, $1 \in A$, $2 \in A$, $3 \in A$. But $7 \notin A$, $6 \notin A$, $13 \notin A$, etc.

Sometimes sets are elements of another set. For instance, the four suits of playing cards are clubs, spades, diamonds, and hearts. Each of these suits is a set consisting of thirteen playing cards. Thus, we can discuss the set of clubs, the set of spades, the set of diamonds, and the set of hearts. The set whose elements are the sets clubs, spades, diamonds, and hearts is a set of sets. That is, it is a set whose elements are sets. For another example, consider the set consisting of all of the teams in the National Football League. Since each individual team is a set of players, the set of all teams in the NFL is a set of sets.

How would you describe a set to someone? Several ways might come to mind. To standardize the procedure of describing sets, the following two methods are used:

1. List all of the elements.

The mathematical convention for listing is to enclose the elements between braces and separate the elements from each other with commas. For example, the set of counting numbers from one to ten, inclusive, is written

$$\{1, 2, 3, 4, 5, 6, 7, 8, 9, 10\}$$

We can extend this *listing* or *roster technique* to large sets if a pattern exists among the elements. This is done by use of the ellipsis, written "..." and read "and so forth." So, to describe the set of all counting numbers from one to one hundred, inclusive, we write

$$\{1, 2, 3, \ldots, 100\}$$

To describe the set of all natural numbers, we write

$$\{1, 2, 3, \ldots\}$$

2. State some property common to each element in the set.

The following example illustrates this method. Think of the letters of the English alphabet as a universal set U. The expression "Each element is one of the first five letters of the English alphabet" characterizes {a, b, c, d, e}. That is, a property that each element possesses is that of being one of the first five letters of the English alphabet. By letting x be a placeholder for the elements of the universal set taken one at a time, and by using the expression "x is one of the first five letters of the English alphabet," we can determine which elements of the universal set are elements of {a, b, c, d, e} in the following way. The expression "x is one of the first five letters of the English alphabet" is true when x is replaced by a. It is also true if x is replaced by b, and similarly for c, d, and e. But, if we pick any other letter of the universal set—such as k—then the expression "x is one of the first five letters of the English alphabet" is false. Suppose we let S(x) represent the expression "x is one of the first five letters of the English alphabet." Then, by enclosing the defining rule, S(x), between set braces, we can represent {a, b, c, d, e} by $\{x \in U$ such that $S(x)\}$. For a more convenient notation, the vertical bar (|) is used in place of the phrase *such that.* Using the same universal set, we can compactly express $\{a, b, c, d, e\}$ by $\{x \in U \mid S(x)\}$.

If $S(x)$ is some assertion regarding x, $\{x \in U \mid S(x)\}$ is the set of all elements x of U such that $S(x)$ is true. This notation can be shortened to $\{x \mid S(x)\}$, since we know x must be a member of a specified universal set. The notation $\{x \mid S(x)\}$ is called *set-builder* notation when $S(x)$ is replaced by an assertion.

Example 2: Let S(x) be the expression "x is the name of a day of the week that begins with *T*." Let U = {Monday, Tuesday, Wednesday, Thursday, Friday, Saturday, Sunday}. Then,

$$\underset{\text{(Set-builder notation)}}{\{x \mid S(x)\}} = \underset{\text{(Roster notation)}}{\{\text{Tuesday, Thursday}\}}$$

Note that a set does not necessarily possess the property common to each of its elements. The common property of each element of $\{a, b, c, d, e\}$ is that of being one of the first five letters of the English alphabet. We see that the set itself, considered as a single entity, does not have this property. That is, $\{a, b, c, d, e\}$ considered as a totality (set) is not one of the first five letters of the English alphabet.

Exercises 1.1

1. State whether or not each of the following is a well-defined set.

a) The states in the U.S. whose names begin with the letter I.
b) The vegetables that taste good.
c) The students who enjoy bicycling.
d) A number that when added to 37 gives 52.
e) The students who like to study mathematics.

2. Give four examples of well-defined sets and four examples of sets that are not well defined.

3. True or False.

a) $6 \in \{1, 2, 3, \ldots, 48\}$
b) $24 \notin \{x \mid x \text{ is a multiple of } 2\}$

c) orange $\in \{y \mid y \text{ is a color}\}$
d) Ford $\notin$ {Buick, Dodge, Chevrolet}
e) $\{a\} \in \{a, b, c\}$
f) $\{a\} \in \{\{a\}, \{b\}, \{c\}\}$
g) $a \in \{\{a\}, \{b\}, \{c\}\}$
h) $p \subset \{\{r, s\}, \{p\}\}$

4. Let A denote the set of all objects that fly. Let B denote the set of all animals. Name an object x such that,
a) $x \in A$ but $x \notin B$
b) $x \in A$ and $x \in B$
c) $x \notin A$ and $x \notin B$
d) $x \notin A$ but $x \in B$

5. Describe the following sets using roster notation.
a) The colors in the United States flag.
b) The natural numbers that are multiples of 6.
c) The months of the year.
d) The counting numbers from 14 to 36, inclusive.
e) The divisors of 8.
f) The different letters used in the word *aluminum*.

6. Name a universal set for each of the following sets.
a) A = {Earth, Mars, Pluto, Mercury, Jupiter}
b) $A = \{x \mid x \text{ is an animal in the San Diego zoo}\}$
c) $A = \{y \mid y \text{ is a person over six feet tall}\}$
d) A = {ruby, diamond, sapphire}
e) A = {steak, potatoes, cake}

7. For each of the following sets name two sets which might possibly serve as universal sets.
a) $A = \{1, 2, 3, 4, 5\}$
b) $A = \{2, 4, 6, \ldots\}$
c) A = {Los Angeles, San Francisco, Sacramento, San Diego}
d) A = {trout, tuna, catfish, salmon, bass}

8. In each problem below, there are two sets given. Give a set that can serve as a universal set for both sets simultaneously.
a) $A = \{1, 3, 5, \ldots\}$ $B = \{2, 4, 6, \ldots\}$
b) A = {rose, tulip, lily} B = {pecan, walnut, peanut}
c) $A = \{3, 6, 9, \ldots\}$ $B = \{8, 16, 24, \ldots\}$

9. Let $A = \{x \mid x$ is a state in the United States whose

name begins with an A}. List the elements in A if the universal set is:

a) The set of all states in the United States.

b) The set of all states whose names end with the letter a.

c) The set of all states west of the Mississippi River.

d) The set of all states north of the 48th parallel.

10. Give a universal set for each of the following sets and then describe the given set using set-builder notation.

a) {5, 10, 15, 20}

b) {ham, bacon, sausage}

c) {3, 4, 5, . . . , 34}

d) {knife, fork, spoon}

e) {Oregon, Ohio, Oklahoma}

f) {c, o, 1, e, g}

g) {Lake Erie, Lake Huron, Lake Michigan, Lake Superior, Lake Ontario}

h) {4}

i) {1, 2, 3, . . .}

j) {2, 4, 6, 8, . . .}

11. Let U = {0, 1, 2, 3, 4, 5, 6, 7, 8, 9}, A = {3, 4, 6, 7}, and B = {5, 1, 3, 9, 7}. List the elements of the following sets.

a) {x | x is in set A or set B or both A and B}

b) {x | x is in both set A and set B}

c) {x | x is in set A or set B but not in both A and B}

12. Let A be a set. What is the difference between the set A and {A}?

13. Let U = {{a, b}, {1, 2,}, {#, &, $}, {pearl}}. Let S($x$) be "$x$ is a set which has exactly 2 elements in it." Let A = {$x \in$ U | S(x)}. Describe A using roster notation.

Notice that even though A has the property that is common to each of its elements, it is not an element of itself, i.e., A $\notin$ A.

14. Let U be the set consisting of the letters of the English alphabet. Let the expression "x is one of the first five letters of the English alphabet" be S(x). Then,

$$\{x \mid S(x)\} = \{a, b, c, d, e\}$$

Discuss the symbol "x" used as a placeholder in the expression S(x) and the symbol "x" as a member of U.

15. Let U = $\{x \mid x$ is a word in the English language$\}$. Consider A = {apple, moon, shoe}.

a) What is a property common to each of the elements of A?

b) What is a property common to each of the elements of A and only the elements of A?

Section 2
Relatives

We are now in possession of some fundamental terminology and symbolism in mathematics. Equipped with such interesting and basic concepts, we are ready to adventure further into the mathematical arena. Before we proceed, however, we pause to mention a few considerations regarding our procedure for further development.

To many people, mathematics seems to consist of concepts related to each other in some mystical way. These concepts and their relationships, however, do not come to earth by some mysterious process, to be collected by only a few people with special truth-catcher antennas. In developing mathematical relationships and concepts, we find that there is normally a natural sequence of events involved. Most of the ideas in this text are developed in response to the natural questions that might arise in considering existing bits of information. Of course, the term *natural* is rather vague, and what seems natural to the authors might seem weird to others, and vice versa. However, from our experience in mathematics education, we believe that you will find the questions to be natural and will recognize the pattern of development that they reveal.

To make the development of ideas flow easily for you, the importance of understanding given definitions cannot be overemphasized. Definitions are to mathematics what mortar is to the bricklayer. They are the cement that holds mathematical structures together. In this same vein, we must mention the importance of becoming comfortable with new symbols that are introduced. These symbols are our mathematical language, and it is in the symbols of mathematics that the power of mathematics lies.

Relationships of Sets

Having considered the nature of sets, we now consider the question, "What are some of the relationships that may or may not exist between sets?" Given two sets, we have three basic relationships: one set may be a portion of the other set, the sets may be exactly the same, or the sets may have the same number of elements but not exactly the same elements.

Subsets

Suppose that set A is a portion of set B. This means that every element in A is also an element of B. Also, if every element in A is an element of B, then A is a portion of B. Since A is a set and B is a set, *subset* is a natural term to replace *portion* in the preceding discussion.

Definition: For sets A and B, A is a *subset* of B if and only if every element in A is an element of B. We write $A \subset B$ or $B \supset A$ (read "A is a subset of B"). If A is not a subset of B, we write $A \not\subset B$.

The phrase *if and only if* might seem stilted or strange to you. Therefore, we will restate the above definition so that it embodies the idea of the *if and only if* phrase: If set A is a subset of set B, then every element in set A is also an element of Set B and if every element of set A is also an element of set B, then set A is a subset of set B. When the phrase *if and only if* is used, it will have the same meaning illustrated for the definition above.

Example 3: Let $U = \{1, 2, 3, \ldots, 10\}$, $A = \{1, 2\}$ and $B = \{1, 2, 3\}$. Then $A \subset B$, since 1 and 2 are elements of A and are also elements of B.

Example 4: Let $U = \{x \mid x \text{ is a car}\}$, $A = \{x \mid x \text{ is a Ford}\}$ and $B = \{x \mid x \text{ is a Ford manufactured since 1960}\}$. Then $B \subset A$.

To show that a set A is not a subset of a set B, we must exhibit an element in A that is not an element of B. Suppose $U = \{1, 2, 3, 4, 5, 6\}$, $A = \{1, 4\}$, and $B = \{1, 2, 3\}$. Then A is not a subset of B since $4 \in A$ and $4 \notin B$. Hence, we write $A \not\subset B$.

It is obvious by the definition that any set is a subset of itself.

Keep the meaning of the symbols $\in$ and $\subset$ clear. The symbol $\in$ expresses the relation between a set and any one of its members. The symbol $\subset$ expresses a relation between a set and one of its subsets. If $A = \{a, b, c\}$, then $a \in \{a, b, c\}$, but $a \not\subset \{a, b, c\}$ although $\{a\} \subset \{a, b, c\}$ while $\{a\} \notin \{a, b, c\}$.

Equal Sets

For two sets to be equal they must be identical. This notion is expressed in the following definition.

Definition: Two sets, A and B, are said to be *equal* if and only if $A \subset B$ and $B \subset A$. Symbolically, we write $A = B$.

$A = \{\#, \&\}$ and $B = \{\&, \#\}$ are equal sets since $A \subset B$ and $B \subset A$. We note that the order in which the elements of a set are listed does not matter.

The sets $A = \{1, 1, 2\}$ and $B = \{1, 2\}$ are such that $\{1, 1, 2\} \subset \{1, 2\}$ and $\{1, 2\} \subset \{1, 1, 2\}$. Thus, $A = B$. Hence, we do not gain anything by listing an element in a set twice. Listing a given element more than once in a set is not desirable notation.

For sets A and B, if $A \subset B$ and $A \neq B$ (read, "A is not equal to B"), we say that A is a *proper subset* of B. Let $A = \{7, 8, 9\}$ and $B = \{4, 5, 6, 7, 8, 9\}$. Then $A \subset B$, but $A \neq B$. Therefore, A is a proper subset of B.

Equivalent Sets

Suppose you have just returned from a shopping trip with an "assemble-it-yourself gadget." You spread the parts out on the floor and notice that according to the instructions there are six bolts, which need to be inserted and fastened in six holes. You immediately look for the six bolts that will pair with the six holes. Hopefully, they will all be in the kit. If you do not have a bolt for some particular hole, you will probably find yourself in a state of emotional disequilibrium; i.e., mad.

Pairing is a fundamental notion in mathematics. Note that the set of bolt holes and the set of bolts (if they are all there) are related to each other so that you are able to form a one-to-one correspondence between the elements of the two sets. Similarly, we can match the elements of the set of names of states in the United States with the elements of the set of names of capitals of states in the United States. These two sets are obviously not the same sets, yet they can be paired element with element.

Definition: Two sets, A and B, are in *one-to-one correspondence* if and only if each element of A can be paired with exactly one element of B and each element of B can be paired with exactly one element of A. The phrase *one-to-one correspondence* may be written *1-1 correspondence.*

All the different 1-1 correspondences for the sets {a, b, c} and {#, %, *} are given in Figure 1.6. Each one of these six matchings demonstrates a 1-1 correspondence.

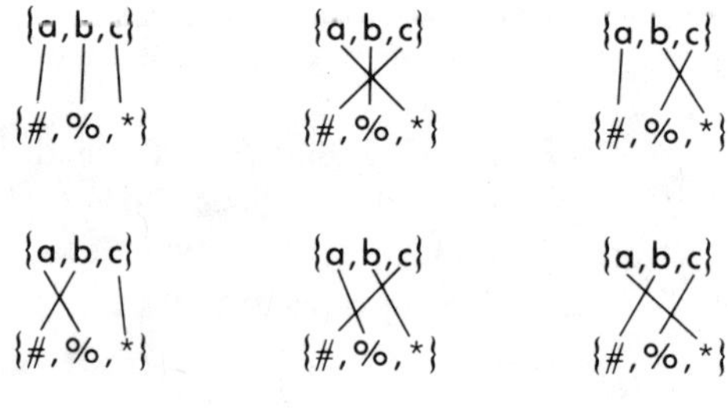

Figure 1.6

Definition: Two sets, A and B, are *equivalent* if and only if there is a one-to-one correspondence between set A and set B. We write this as A ≡ B.

Notice that a set is always equivalent to itself. If A = {a, e, i, o, u}, then the correspondence

$$\begin{array}{ccccc} \{a, & e, & i, & o, & u\} \\ | & | & | & | & | \\ \{a, & e, & i, & o, & u\} \end{array}$$

demonstrates that A is equivalent to A. Also, it is clear that any 1-1 correspondence between two sets, A and B, shows that A ≡ B as well as B ≡ A.

Let us consider the process involved in getting a pair of eyeglasses. First, the left and right eyes are matched to correctional instructions: a left-eye correctional instruction (L.E.C.I.) for the left eye and a right-eye correctional instruction (R.E.C.I.) for the right eye. The left lens is then ground following the instructions for the left eye and the right lens is ground following the instructions for the right eye. The left lens is then matched to the left eye and the right lens to the right eye. We illustrate this relationship as follows:

$$\begin{array}{lccc} \text{First,} & A = & \{\text{Left eye,} & \text{Right eye}\} \\ & & | & | \\ & B = & \{\text{L.E.C.I.,} & \text{R.E.C.I.}\} \\ \text{and then,} & B = & \{\text{L.E.C.I.,} & \text{R.E.C.I.}\} \\ & & | & | \\ & C = & \{\text{Left lens,} & \text{Right lens}\} \\ \text{from which we get,} & A = & \{\text{Left eye,} & \text{Right eye}\} \\ & & | & | \\ & C = & \{\text{Left lens,} & \text{Right lens}\} \end{array}$$

The preceding illustration points up a very useful concept in mathematics: if A is equivalent to B and B is equivalent to C, then A is equivalent to C.

The words *equal* and *equivalent* have meanings that are quite different. We note that all equal sets are equivalent, but not all equivalent sets are equal.

Section 3
Something for Nothing

We recall that for each set under consideration there is an associated universal set and some explicit criterion that, when applied to the universal set, divides the universal set into those elements in the set under consideration and those elements in the set not under consideration. Consider the universal set $U = \{1, 3, 5, 7, 9, 11, 13\}$, and let $S(x)$ be the expression "x is an even number between one and thirteen." We see that there is no element of U that makes $S(x)$ true. $S(x)$ separates U into U and a set with no elements. The elements not in $\{x \in U \mid S(x)\}$ happen to be all of the elements of U. Hence, $\{x \in U \mid S(x)\}$ has no elements.

Definition: A set is called an *empty set* if and only if the set contains no elements. We designate an empty set by the symbol ϕ, or by $\{\ \}$.

We note some interesting properties of the empty set.

1. If A is any set, then $\phi \subset A$.

We noted earlier that to show that a set B is not a subset of a given set A we must exhibit an element in B that is not in A. Let A be any set. Can you find an element of ϕ that is not in A? Obviously not (even if you are a magician) since ϕ has no elements. Hence, we cannot say that ϕ is not a subset of A. We have one other choice since sets are either subsets of each other or they are not subsets of each other. That choice is that ϕ is a subset of A.

Perhaps you have not encountered this type of argument before. If not, settle back, read it again, and double your enjoyment of the subtleties involved.

The authors have found that some students have come to grief by confusing the fact that the empty set is a subset of any set with the statement "ϕ is an element of every set." It is clear that ϕ is not an element of every set. For instance, A = {1, 2, 3} is a set that does not have ϕ as an element, i.e., $\phi \notin A$. On the other hand, if we let B = {1, 2, 3, ϕ}, then $\phi \in B$ and $\phi \subset B$.

2. There is only one empty set.

Suppose there exists more than one empty set. That is, suppose both ϕ and ψ are empty sets. From our set discussion we found that $\phi \subset A$ for any set A. Let $A = \psi$. Then we have that $\phi \subset \psi$. But then, if we let $A = \phi$ and remember that ψ is empty, this means $\psi \subset \phi$. Hence, $\phi \subset \psi$ and $\psi \subset \phi$, which means $\phi = \psi$ from our definition of two sets being equal. We conclude that there is only one empty set.

Referring back to our previous section, we recall discussing one-to-one correspondences and equivalent sets. It becomes awkward to speak of a one-to-one correspondence between sets A and B when they are both empty. So, if A and B both represent the empty set, we define A to be equivalent to B.

Exercises 1.2, 3

1. Construct seven different 1-1 correspondences between A = {1, 2, 3, 4} and B = {p, q, r, s}.

2. Let A = {e, l, i, a, v} E = {u, n, r, t, s, e}
B = {t, e, v, s, n, u} F = {e, t, m, r, s}
C = {s, t, r, e} G = {v, l, e, i}
D = {m, i, r, s, l, a} H = {l, i, r, a, s, m}

Using the above sets, decide whether each of the following is true or false.

a) $D \equiv H$	d) $C \not\equiv G$	g) $D \equiv F$	j) $F = A$
b) $G \subset A$	e) $E = H$	h) $H \not\subset D$	k) $B \not\equiv H$
c) $E \neq B$	f) $\phi \subset A$	i) $E \not\subset C$	l) $C \subset F$

3. Which of the following sets are empty sets?
a) All birds that live underwater.
b) All babies that cry.
c) All houses that are painted blue.
d) All live mammals that cannot breathe.
e) All countries of the world that grow tea.

4. List all the subsets for each set that follows:
a) A = {1}
b) B = {1, 2}
c) C = {1, 2, 3}
d) D = {1, 2, 3, 4}

Can you make a conjecture as to the relation between the number of elements a set has and the number of subsets the set has? To help you express the relation, we note the following convention. A short way to express $2 \cdot 2 \cdot 2$ is 2^3. In general, we write 2^n for $\underbrace{2 \cdot 2 \cdot 2 \ldots 2.}_{n \text{ times}}$

5. Let A = the set of all vowels.
B = the set of different letters in the phrase, "Listen here, Sam, be still!"
C = the set of different letters in the word *establishmentarianism*.
D = the set of different letters in "such ignorance."

Using the above-listed sets, answer the following:
a) Which of the above sets are equal?
b) Which of the above sets are equivalent?
c) State all the subset relationships.
d) State all proper subset relationships.

6. List all the subsets of the empty set. How many proper subsets does the empty set have?

7. True or False?
a) $\phi \in \phi$
b) $\phi = \{\ \}$
c) $\{\phi\} = \phi$
d) $\phi \subset \phi$
e) $\phi \subset \{\phi\}$
f) $\phi \in \{\phi\}$
g) $\phi = \{0\}$
h) $\{\phi\} = 0$
i) $\phi \equiv \phi$
j) $0 \subset \{0\}$

8. Let A = the set of all black dogs.
B = the set of all stray dogs in Santa Fe.

Using the above, state in words what each of the following expressions says symbolically.
a) $B \subset A$
b) $A \not\subset B$

9. Let U = {0, 1, 2, 3, 4, 5, 6, 7, 8, 9}. Let $S(x)$ be the

expression "x is a digit." Describe the two sets that are formed by applying $S(x)$ to the given universal set, U.

10. In each of the following, how many different 1-1 correspondences can be constructed between the two sets given?

a) A = {1} and B = {a}
b) A = {1, 2} and B = {a, b}
c) A = {1, 2, 3} and B = {a, b, c}
d) A = {1, 2, 3, 4} and B = {a, b, c, d}

Can you make a conjecture as to the relationship between the number of elements that two equivalent sets possess and the number of 1-1 correspondences that can be constructed between the two sets?

11. How many different ways can 6 people be seated in 6 chairs? (How many 1-1 correspondences exist between the set of 6 chairs and the set of 6 people?)

12. How many different ways can 5 books be arranged on a bookshelf? (How many 1-1 correspondences exist between the set of 5 books and the set of 5 positions on the bookshelf?)

13. In each of the following, decide whether the sets given are equivalent or not. Then write A ≡ B or A ≢ B.

a) A = {5, 8, 4, 3} B = {hot, cold, warm, cool}
b) A = {1, 2, 3, . . ., 30} B = {31, 32, 33, . . ., 60}
c) A = {ϕ} B = ϕ
d) A = {2, 4, 6, . . ., 100} B = {1, 3, 5, . . ., 101}
e) A = {18, 21, 24, . . ., 72} B = {5, 10, 15, . . ., 95}

14. Construct a 1-1 correspondence between A = {a, b, c} and B = {{a}, {b}, {c}}. A set which consists of only one element is called a *singleton set*. Thus, {a}, {b}, {c} are all examples of singleton sets. Hence, {{a}, {b}, {c}} is a set whose elements are singleton sets.

Is {ϕ} a singleton set? Is {{3}} a singleton set?

Let A = {1,2}. The set B = {{1}, {2}} is in 1-1 correspondence with A.

In a manner similar to that above, construct a set that is in 1-1 correspondence with each of the following sets so that its elements are singleton sets formed from elements of the given set.

a) A = {\$, ¢, @}
b) A = {orange, yellow}
c) A = {7, 5, 4, 6}
d) A = {*}

15. Let A = $\{\phi, \{1\}, \{2\}\}$; B = $\{1, 2\}$; C = $\{\{1\}, \{1, 2\}, 1, 2\}$. State whether the following are true or false.

a) $\phi \subset A$
b) $\phi \not\subset C$
c) $B \in C$
d) $B \not\subset C$
e) $\phi \in A$
f) $A \not\subset C$

16. Let A = $\{a, b\}$

a) Construct a set B such that $A \in B$.
b) Construct a set C such that $A \subset C$.
c) Construct a set D such that $A \in D$ and $A \subset D$.

17. List all the subsets of the set $\{a, \{b, c\}\}$.

Section 4

Something Old, Something New

Sets Formed from Subsets of a Given Set

Let A be a set. By pulling together the notion of subset and the idea that sets can be elements of sets, we can see that it is reasonable to consider a set that contains all or some of the subsets of A. To be explicit, suppose we let $A = \{1, 2\}$. It is clear that $\{1\}$ and $\{2\}$ are subsets of A. Recall that ϕ is a subset of any set; hence, it is a subset of A. Also, any set is a subset of itself so that $A \subset A$. Thus, the subsets of A are ϕ, $\{1\}$, $\{2\}$, and $\{1, 2\}$. We can list them as the set $\{\phi, \{1\}, \{2\}, \{1, 2\}\}$. Thus, we have a set that has been built from the set A. We could have many other sets that have some of the subsets of A as elements: $\{\phi, \{1, 2\}\}$, $\{\{1\}, \{2\}\}$, $\{\phi, \{1\}, \{2\}\}$, and others. The set that has for its elements all of the subsets of a given set A is of particular interest in mathematics and is called the *power set* of A, written $\mathcal{P}(A)$.

Example 5: If $A = \{a, b, c\}$, then $\mathcal{P}(A) = \{\phi, A, \{a\}, \{b\}, \{c\}, \{a,b\}, \{a,c\}, \{b,c\}\}$. Notice that $b \in A$, $\{b\} \in \mathcal{P}(A)$, $\{b\} \subset A$, and $\{\{b\}\} \subset \mathcal{P}(A)$.

Note that for a set A, $\mathcal{P}(A)$ has the property that $\phi \in \mathcal{P}(A)$ and $\phi \subset \mathcal{P}(A)$.

Sets Formed by Operations on Given Sets

We have four basic arithmetic operations. These operations are addition, subtraction, multiplication, and division:

$$8 + 4 = 12 \qquad 8 \cdot 4 = 32$$
$$8 - 4 = 4 \qquad 8 \div 4 = 2$$

To try to get an intuitive notion of what we mean by an operation, we will consider a common property of these basic operations. We note that in each case we take two numbers, combine them according to a rule, and get a resultant number. The idea of *operation* can be intuitively described as a rule or method for associating a pair of numbers (or objects) with another number (or object) of the same kind. Later we will extend the idea of operation to operations on single objects. Perhaps it would be good to make a distinction between the idea of "relation," which we have already discussed, and the idea of "operation" as these ideas relate to sets. A relation exists between two sets if it is apparent that the sets are associated in some manner. An operation on sets is a rule or technique for associating a pair of sets with another set.

Given two sets A and B, which are subsets of the same universal set, what are some sets that naturally arise from these sets?

Intersection

Suppose an employer wishes to hire a secretary. After he has the information from the interviews with applicants for the position, he uses the following procedure to help him select the secretary-to-be. He considers only those who can type more than 75 words per minute and compiles a list of applicants who meet this requirement. He also wants someone who can take shorthand and compiles a list of applicants who meet this requirement. The final list he compiles is composed of the names that appear on both previous lists.

This example illustrates the operation of intersection. Similarly, when we talk about the intersection of two streets, we are talking about the area that both streets have in common. Also, when we speak of the intersection of two lines, we are referring to the point both lines have in common. The operation of intersection associates with a pair of sets the set consisting of all objects that the pair of sets have in common.

Definition: Let A and B be sets. The *intersection* of A and B is the set of all elements that are in both A and B. Symbolically written,

$$A \cap B = \{x \mid x \in A \text{ and } x \in B\}.$$

Notice that this definition means:

$$x \in A \cap B \text{ if and only if } x \in A \text{ and } x \in B$$

Suppose we have two buckets of objects. We decide to make up a third bucket of objects according to the following rule: for an object to be placed in the third bucket, it must appear in both of the other buckets; for example, let buckets A and B have as elements the letters of the English alphabet as indicated in Figure 1.7; then, according to our rule, bucket C has the letters c, d, and e.

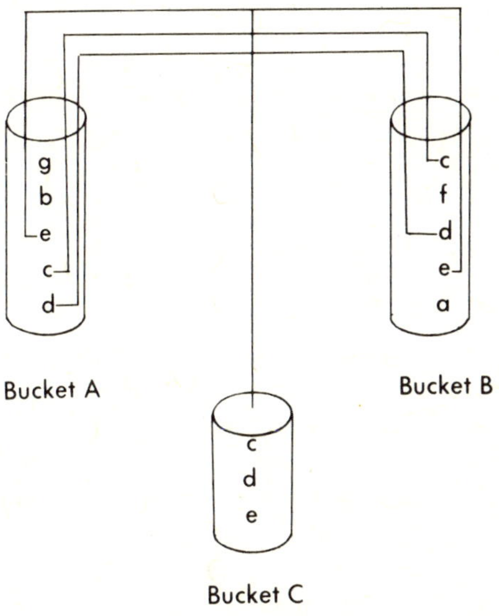

Figure 1.7

Let's look at another example. Let A = {1, 2, 3, 4, 5, 6} and B = {3, 4, 7, 8}. Then, A ∩ B = {3, 4}. Notice that we do not write A ∩ B = {3, 3, 4, 4} because, as previously stated, listing an element twice is poor notation.

Let A = {a, b, c} and B = {e, f, g}. Then A ∩ B = ϕ When two sets, A and B, have no elements in common (A ∩ B = ϕ), they are *disjoint sets*.

Union

Suppose a college requires that students must have a four-hour laboratory course in life science to graduate. The students can fulfill this requirement by taking either a four-hour course in zoology or a four-hour course in botany. The list of the students who meet this requirement is composed of those students who have completed four hours of zoology along with those who have completed four hours of botany. This example illustrates the operation of union.

Similarly, the carpenter and mill workers' federation is made up of all of the individuals who consider themselves to be either carpenters or mill workers or both. The operation of union associates with a pair of sets that set consisting of all objects that are in one set or the other set (or both).

Definition: Let A and B be sets. The *union* of A and B is the set of all elements that are in set A or in set B (or in both). Symbolically we write,

$$A \cup B = \{x \mid x \in A \text{ or } x \in B\}.$$

Notice that this is equivalent to:

$$x \in A \cup B \text{ if and only if } x \in A \text{ or } x \in B.$$

A comment regarding the use of the word *or* in the definition of union is in order. The word *or* is used in two different ways. The *exclusive* use of *or* occurs when we mean *one or the other but not both*. For example, if you tell someone, "I am going to the ball game or to the library tonight," you have used or in the exclusive sense. The *inclusive* use of the word *or* occurs when we mean *one or the other or both*. In the example regarding the four-hour life science requirement for graduation, it was stated that a student had to have four hours of zoology or four hours of botany to graduate. Yet it is not the case that having both courses would endanger graduation. In our definition of A ∪ B one should read $x \in A$ or $x \in B$ to mean *x is an element of A or x is an element of B or x is an element of both A and B.*

To illustrate the union of two sets, we take bucket A and bucket B and form a third bucket, bucket C, by putting in bucket C all those objects that occur either in bucket A or in bucket B as well as those objects that may be in both buckets. Figure 1.8 illustrates this process.

For another example, let A = {a, b, 1, 2, 7, 3} and B = {a, 1, &, 3, m, z}. Then A ∪ B = {a, b, m, z, 1, 2, 3, 7, &}.

Complementation

The operations of intersection and union are called *binary operations,* since these operations "operate" on two sets.

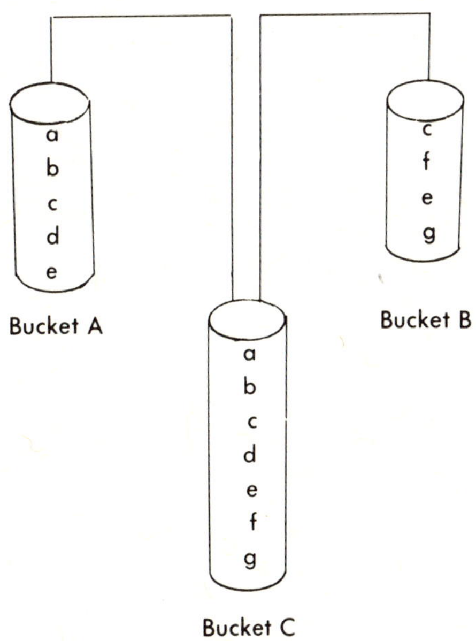

Figure 1.8

Another operation is a *unary* or *singular operation*. This operation assigns one set to another set instead of assigning a pair of sets to another set. *Complementation* is a unary operation.

As we have seen from the previous material, every set is associated with some universal set. In considering the operation of complementation, it will be quite clear that an associated universal set must be specified if this operation is to make any sense.

Definition: Let A be a subset of a universal set U. A set is the *complement of A in U* if and only if it consists of all elements of U that are not in A. The complement of A in U is denoted by $\bar{A}$, read "not A" or "the complement of A" or "the negation of A". Writing the complement symbol, –, over a set expression expresses the complement of the expression under the bar. For example, the complement of $A \cap B$ is $\overline{A \cap B}$.

Consider the buckets again as they relate to the operation of complementation. Let bucket U be the bucket that has as its elements all of the vowels in the English alphabet, and let bucket A have as its elements the letters a and u. Bucket $\bar{A}$ would have as its elements the letters o, e, and i since these are the letters that are in bucket U and not in bucket A. Figure 1.9 illustrates this example.

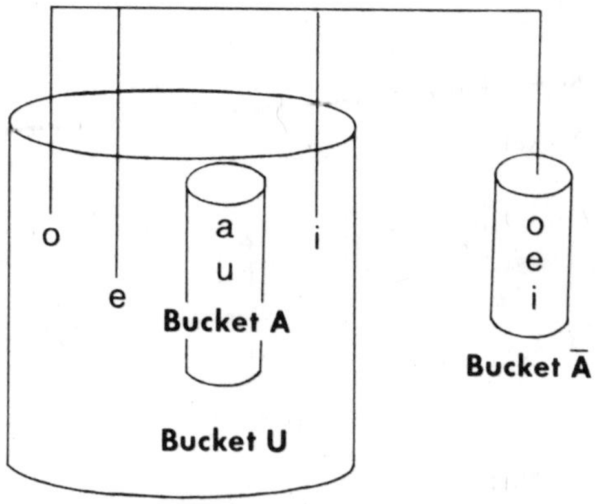

Figure 1.9

For another example, let $U = \{1, 2, 3, 4, 5, 6, 7, 8, 9\}$ and $A = \{1, 3, 5, 7, 9\}$. Then $\bar{A} = \{2, 4, 6, 8\}$.

Relative Complement

A binary operation closely associated to complementation is the *relative complement.*

Definition: Let A and B be sets. A set is the *relative complement of B with respect to A* if and only if it consists of all those elements of A that are not in B. The relative complement of B with respect to A is denoted $A - B$. Symbolically, we write

$$A - B = \{x \mid x \in A \text{ and } x \notin B\}$$

Notice that according to the definition, $A - B = A \cap \bar{B}$. For example, let $U = \{a, b, c, d, e, f, g\}$, $A = \{a, b, c, d\}$, and $B = \{c, d, e, f\}$. Then, $A - B = \{a, b\}$.

Sets Formed by Extended Application of Operations

What happens if we have three sets that are subsets of the same universal set? Can we use our binary or unary operations to produce a new set from the three sets? Suppose we try.

Let $U = \{1, 2, 3, 4, 5, 6, 7, 8, 9\}$, $A = \{1, 2, 3\}$, $B = \{2, 3, 4\}$, and $C = \{4, 5, 8\}$. Consider an expression of the form $A \cap B \cup C$. Since our operations, $\cap$ and $\cup$, are defined as binary, we can only work with two sets at a time. Thus, $A \cap B \cup C$ has two possible interpretations. First, let's consider $A \cap B$ and its union with C. $A \cap B = \{2, 3\}$. So, $\{2, 3\} \cup C = \{2, 3, 4, 5, 8\}$. Now let's consider $B \cup C$ and its intersection with A. $B \cup C = \{2, 3, 4, 5, 8\}$ and $\{2, 3, 4, 5, 8\} \cap A = \{2, 3\}$. Since the resulting sets are different, we see that $(A \cap B) \cup C \neq A \cap (B \cup C)$. Thus, the order of operations in working with $A \cap B \cup C$ is important. So that an expression will not be ambiguous, we use parentheses to give us procedural instructions for working with it. The expression $(A \cap B) \cup C$ tells us first to intersect A and B and then union this result with C.

In English, we punctuate a sentence with commas to

keep our meaning clear. For instance, removing the commas in the sentence, "Do you see, Mary?" would give us, "Do you see Mary?" Parentheses are mathematical punctuation. To keep a given expression from being ambiguous, we use parentheses to make our meaning precise.

Example 6: Let U = {1, 2, 3, 4, 5, 6, 7, 8, 9}, A = {2, 3, 4, 7}, B = {3, 4, 5, 6}, and C = {5, 6, 7, 8}. What are the elements in $(A \cup B) \cap \bar{C}$? The parentheses tell us to find $A \cup B$, which is {2, 3, 4, 5, 6, 7}. We then find $\bar{C}$ = {1, 2, 3, 4, 9}. Finally, $(A \cup B) \cap \bar{C}$ = {2, 3, 4, 5, 6, 7} ∩ {1, 2, 3, 4, 9} = {2, 3, 4}. So $(A \cup B) \cap \bar{C}$ = {2, 3, 4}.

We might have occasion to need additional symbols for clarification. In the expression $A \cap (B \cup C) \cap D \cup (E \cup F)$, the parentheses are not sufficient to keep the expression from being ambiguous. We use brackets, [ ], for further procedural instructions. For example, $A \cap (B \cup C) \cap D \cup (E \cup F)$ can be written $[A \cap (B \cup C)] \cap [D \cup (E \cup F)]$. In general, we do the operations indicated in parentheses first and then the operations in brackets. If the expression is so complex that parentheses and brackets are not adequate to avoid ambiguity, braces, { }, can be used.

Exercises 1.4

1. Let A = {5, 7, 8, 12, 13}, and let B = {4, 6, 9, 11, 8, 12}.
a) Is $12 \in A \cup B$?
b) Is $12 \in A \cap B$?
c) Describe the set $A \cup B$ by using roster notation.
d) Describe the set $A \cap B$ by using roster notation.

2. Let U = {1, 2, 3, . . ., 13}, A = {3, 4, 5, 6, 7}, and B = {4, 6, 7, 9, 11}. Describe the following sets by using roster notation.

a) $\bar{A}$
b) $\bar{B}$
c) $A - B$
d) $B - A$
e) $U - A$
g) $\bar{A} - U$
h) $\bar{A} \cap \bar{B}$
i) $\overline{A \cap B}$
j) $\bar{A} \cup \bar{B}$
k) $\overline{A \cup B}$
m) $B - (A - B)$
n) $A - (\overline{A \cup B})$
o) $U - \bar{B}$
p) $\bar{\bar{U}}$
q) $\bar{\bar{A}} \cup \bar{\bar{B}}$

f) $U - U$ l) $\overline{\bar{A} \cap \bar{B}}$ r) $\bar{\bar{A}} \cap \bar{\bar{\bar{B}}}$

3. Let U = the set of all living people.
A = the set of all redhaired people.
B = the set of all people who drive Cadillacs.
C = the set of all people who wear glasses.

Express the following statements symbolically by using the symbols U, A, B, C, =, ≠, ∩, ∪, ⊂, ⊄, ¯, and ϕ.

a) Some redhaired people do not wear glasses.
b) All of the people who drive Cadillacs wear glasses.
c) There are no redhaired people who drive Cadillacs.
d) Everyone is either a redhaired person or a driver of a Cadillac.
e) No one drives a Cadillac and does not wear glasses.

4. Suppose $A \cup B = B$ and $A \cap B = \phi$. What do you know about A?

5. Suppose $A \cap B = B$. What do you know about B?

6. Suppose $A \cup B = B$. What do you know about A?

7. Suppose $A \cap B = B$ and $A \subset B$. What do you know about A and B?

8. Suppose $A \cap B = \phi$ and $A \cup B = U$. What do you know about B?

9. Suppose $A \cup B = \phi$. What do you know about A and B?

10. Suppose $A \cap B = U$. What do you know about A and B?

11. Suppose $A \cup \bar{B} = U$. What do you know about A and B?

12. Suppose $A - B = \phi$. What do you know about A and B?

13. Suppose $\overline{A \cup B} = \bar{A} \cup \bar{B}$. What do you know about A and B?

14. What is the union of all of the subsets of the universal set $U = \{1, 2, 3, \ldots, 100\}$?

15. Is there a set A such that $A \cap A = \phi$? If so, what is it?

16. Find the power set, $\mathcal{P}(A)$, for each of the following sets.

a) $A = \{a\}$. How many elements does $\mathcal{P}(A)$ have?
b) $A = \{a, b\}$. How many elements does $\mathcal{P}(A)$ have?

c) A = {a, b, c}. How many elements does $\mathcal{P}$(A) have?
d) A = {a, b, c, d}. How many elements does $\mathcal{P}$(A) have?
Can you make a conjecture as to the relation between the number of elements in a set and the number of elements in the associated power set? (Hint: See Problem 4, Exercises 1.2, 3).

17. Let U = {1, 2, 3, . . . , 100}. Consider all subsets of U and the power sets that can be formed from each of these subsets. What is the intersection of all of these power sets?

18. Let A be any set. Give a formula to express the number of elements the set $\mathcal{P}(A) \cup A$ has.

19. Let A = {1, 2}
a) Find the power set of A; that is, find $\mathcal{P}$(A).
b) Find the power set of $\mathcal{P}$(A); that is, construct the set consisting of all possible subsets of $\mathcal{P}$(A). This set is referred to symbolically as $\mathcal{P}(\mathcal{P}(A))$.

20. Let A = {1, 2, 3, 4, 5}, B = {4, 5, 6, 7}, and C = {7, 8, 9}.
a) Are A and B disjoint?
b) Are B and C disjoint?
c) Are A and C disjoint?
d) Are $A \cap B$ and C disjoint?
e) Do A, B, and C have any elements in common?
f) Are A and $B \cap C$ disjoint?

Three or more sets are said to be disjoint if and only if there is no element common to each and every one of the sets under consideration. Hence, as illustrated above, A and B and C can be disjoint, while A and B are not disjoint.

21. Let A be a set. If you take the complement of A an odd number of times, is the resulting set A or $\bar{A}$? For example, consider $\overline{\overline{\overline{A}}}$. If you take the complement of A an even number of times, is the resulting set A or $\bar{A}$?

22. Let A = {a, b, c}, B = {1, 2, 3, 7}, and C = {b, 3, 9}. State whether the following assertions are true or false.
a) $(A \cap B) \cap C = A \cap (B \cap C)$
b) $A \cup (B - C) = (A \cup B) - (A \cup C)$
c) $(A \cup B) \cup C = A \cup (B \cup C)$
d) $(A - B) - C = A - (B - C)$

e) $A \cup (B \cap C) = (A \cup B) \cup (A \cap C)$
f) $A \cap (B \cup C) = (A \cup B) \cap (A \cup C)$

23. Let $A = \{a, e, i, o, u\}$, $B = \{a, b, c, d, e\}$, $C = \{b, c, d, f, g\}$ and $U = \{a, b, c, d, e, f, g, h, i, j, k, o, u\}$. Describe the following sets by using roster notation.

a) $[(A \cup B) \cap C] - A$
b) $B \cap [C - (A \cap B)]$
c) $(\bar{A} \cap B) - (C \cup B)$
d) $[A \cap (B - A)] \cup \{C - [B \cup (C \cap A)]\}$
e) $[(B - C) \cup A] \cap \{[(\overline{A - \overline{B}}) - B] \cap C\}$

24. For sets A and B, subsets of U, we define the binary operation *symmetric difference*, denoted $A \Delta B$, by

$$A \Delta B = \{x \mid x \in A \cap \bar{B} \text{ or } x \in B \cap \bar{A}\}$$

Let $U = \{1, 2, 3, \ldots, 12\}$, $A = \{1, 3, 4, 5, 7, 9\}$, and $B = \{2, 4, 6, 8, 9\}$. What is $A \Delta B$?

Section 5
Back to the Coloring Book

A mathematician uses diagrams as a device to aid his thinking in finding relationships that exist among mathematical ideas. Venn diagrams* are a means of portraying sets visually.

Many of the operations and relations concerning sets presented in our work so far can be illustrated by use of Venn

*Leonard Euler (1707–1783) introduced a way of thinking about sets in terms of circular regions. John Venn (1834–1923), a British logician, introduced further refinements to the idea by using a rectangular region to portray the universal set and circular regions within the rectangular region to designate the other sets. Since Venn was the first to use circular regions in this general way, these diagrams are known as Venn diagrams.

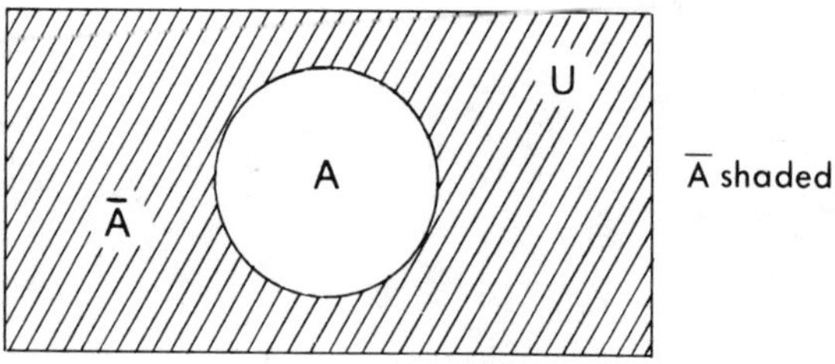

Figure 1.10

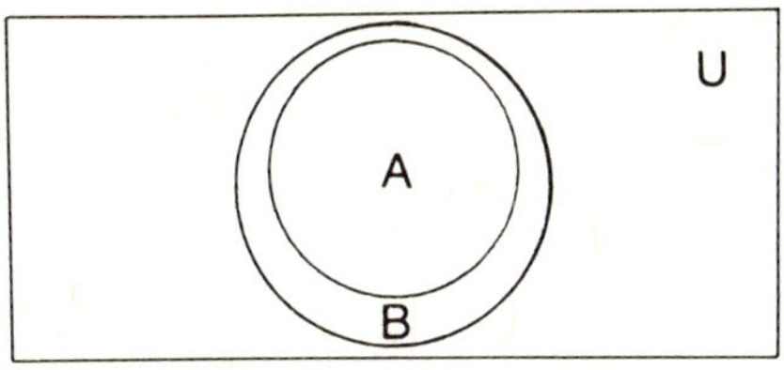

A ⊂ B

Figure 1.11

diagrams. These diagrams portray a universal set and some of its subsets. The universal set U is shown as a rectangular region, and its subsets (A, B, C, etc.) as circular or other regions within the rectangular universal set region.

Perhaps the simplest portrayal is in the case of complement, when we are considering just one subset of the universal set. The complement of set A, designated $\bar{A}$, is shown as the shaded region in Figure 1.10.

In general, if we have two unequal sets, A and B, they may be related in four possible ways, as illustrated in Figures 1.11 through 1.14.

If we do not have any information regarding the relationship of two sets, A and B, and we want to use the pictorial device of Venn diagrams, we use the most general representation to show the relationship of the two sets.

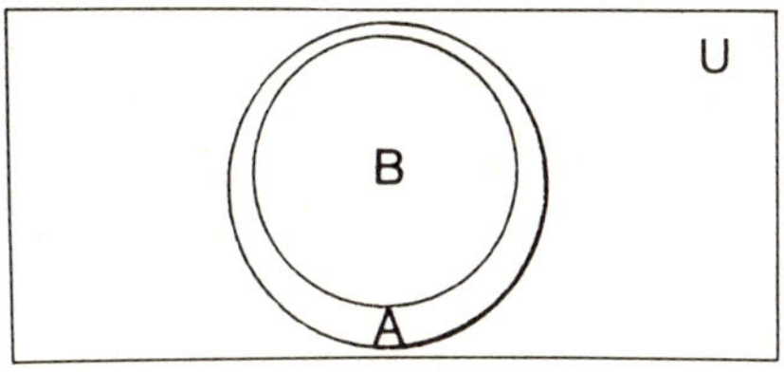

B ⊂ A

Figure 1.12

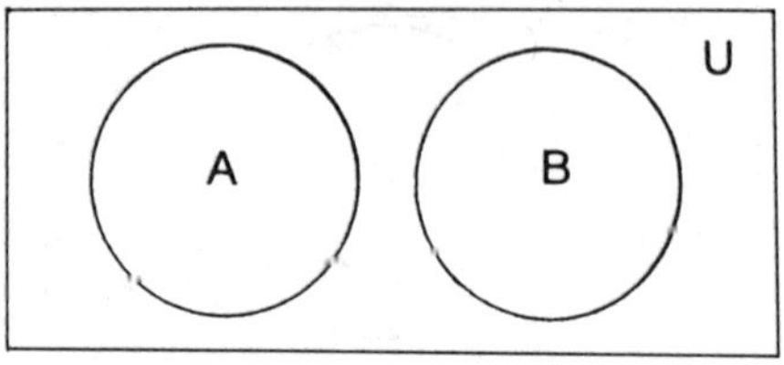

$A \cap B = \phi$

Figure 1.13

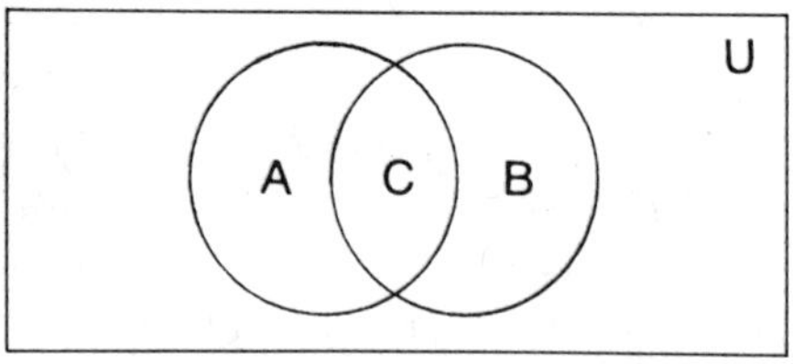

$A \cap B = C$

Figure 1.14

The diagram in Figure 1.15 has four regions, as indicated by the numerals 1, 2, 3, and 4. If we consider region 2 to be empty, then all of A is in B; that is, $A \subset B$ (Figure 1.11). If we consider region 4 to be empty, all of B is in A; that is, $B \subset A$ (Figure 1.12). If we consider region 3 to be empty, there are no elements common to A and B; that is, $A \cap B = \phi$ (Figure 1.13). Hence, we see that Figure 1.14 can be used to represent all these cases. Thus, this diagram is used for the general representation of two sets.

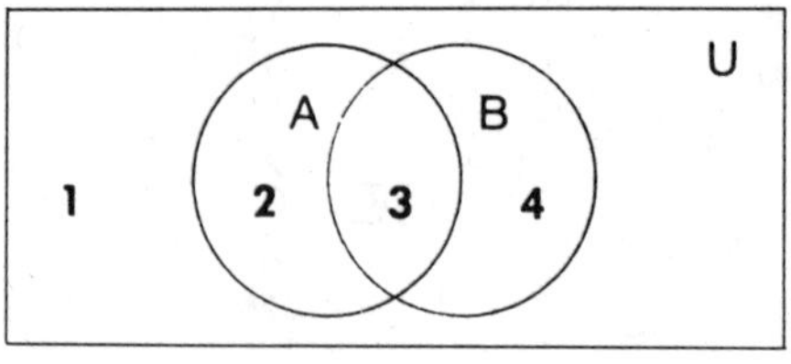

Figure 1.15

By using the general representation of two sets, A and B, we can diagrammatically portray some of the basic operations that we have already discussed.

The intersection of two sets, A and B, is shown as the shaded area in Figure 1.16.

The union is shown as the shaded area in Figure 1.17.

It is sometimes desirable to show more than two sets in a Venn diagram. The general way for representing sets A, B, and C is given in Figure 1.18. We have numbered each

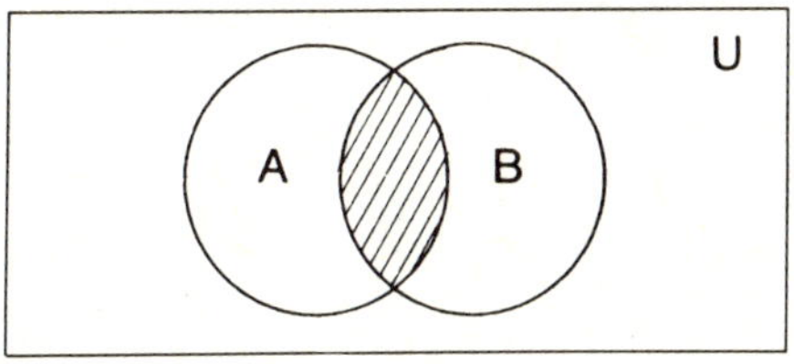

A ∩ B shaded

Figure 1.16

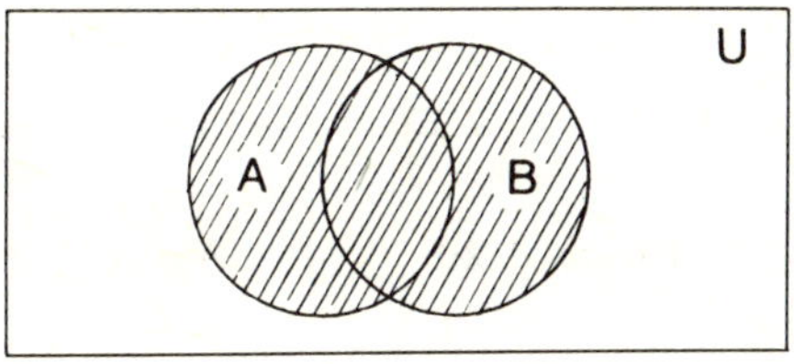

A ∪ B shaded

Figure 1.17

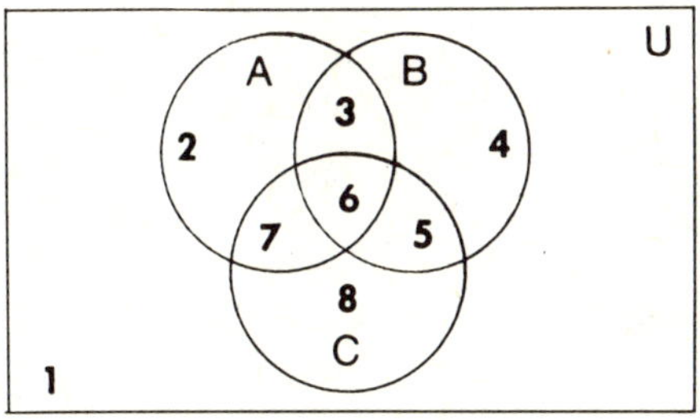

Figure 1.18

separate region with the numerals 1, 2, 3, . . . , 8. A generalized Venn diagram showing more than three subsets of the universal set requires some ingenuity as to ways of representing all possibilities. For most of our work with Venn diagrams we will be dealing with only two or three subsets of a universal set.

Let A, B, and C be sets. How may we use Venn diagrams to portray $(A \cap B) \cup C$? First, shade the area represented by $A \cap B$ (Figure 1.19). This includes regions 3 and 6. Then shade C (Figure 1.20). This includes regions 5, 6, 7, 8. Next, union regions 3 and 6 with regions 5, 6, 7, 8.

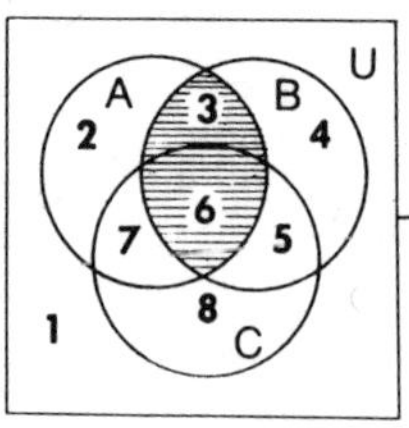

A ∩ B shaded

Figure 1.19

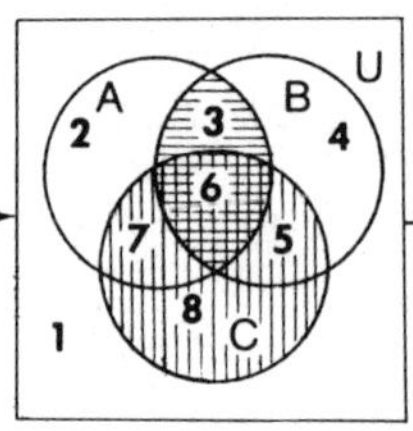

A ∩ B shaded
C shaded

Figure 1.20

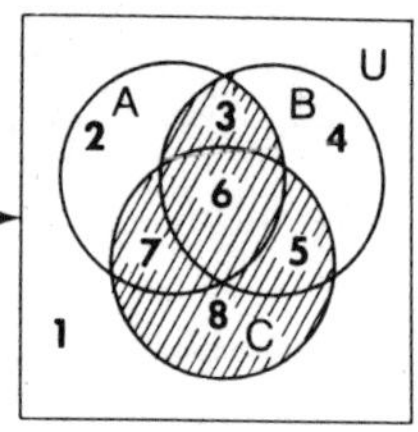

(A ∩ B) ∪ C shaded

Figure 1.21

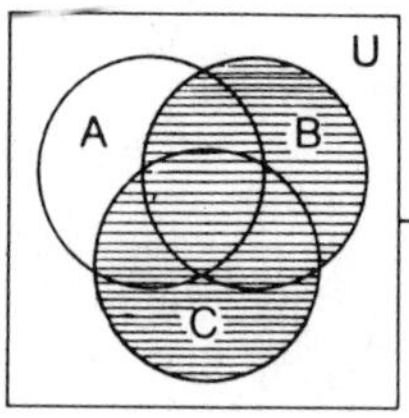

B ∪ C shaded

Figure 1.22

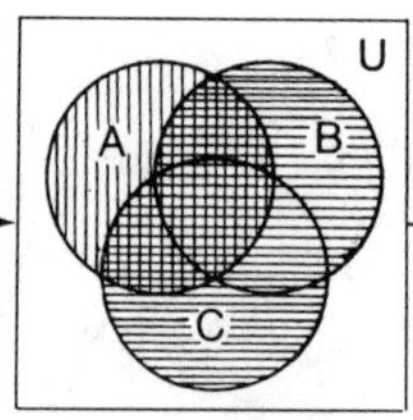

B ∪ C shaded
A shaded

Figure 1.23

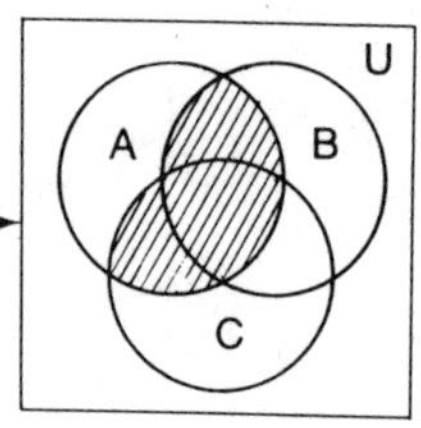

A ∩ (B ∪ C) shaded

Figure 1.24

From the definition of union we see our resultant shaded area includes regions 3, 5, 6, 7, 8 (Figure 1.21).

Similarly, we represent A ∩ (B ∪ C) in Figures 1.22 through 1.24.

There are some interesting types of problems that Venn diagrams can aid in solving. For example, we can apply Venn diagrams to survey, or counting, problems.

Example 7: In a survey of 70 boys participating in athletics, the numbers of athletes participating in various sports were found to be as follows:

football 31
basketball 14
wrestling 22
football and wrestling 7
basketball and wrestling 5
football and basketball 9
football, wrestling, and basketball 4

In the Venn diagram in Figure 1.25, the three sets representing athletes are labeled F, B, and W, for football, basketball, and wrestling, respectively. To find the appropriate numerals to write in the various regions, we begin with the intersection of all three sets, F, B, W. We note that there are 4 students participating in all three sports. There are 9 students in F ∩ B. To find how many are in football and bas-

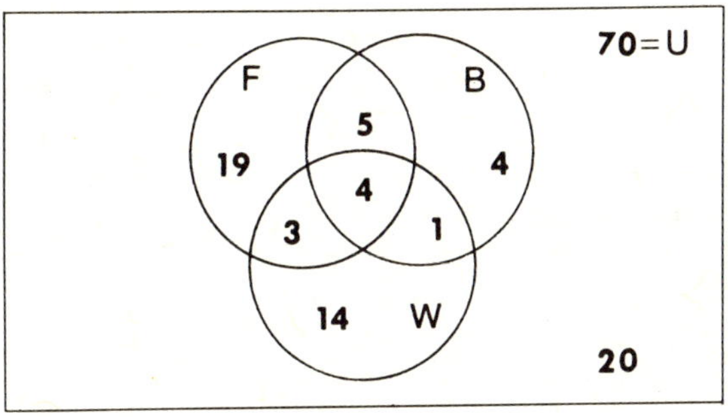

Figure 1.25

ketball but not in wrestling, we subtract 4 from 9 to put 5 in the $(F \cap B) - W$ region. In a similar way, we get 3 to put in the $(F \cap W) - B$ region. We continue the process until all eight regions have the appropriate numerals in them. We note the following:

a) The number of students participating in football or wrestling or both but not in basketball is 36.

b) The number playing basketball only is 4.

c) How many students participated in athletics but not in football, basketball, or wrestling?

To answer this question, we find out how many participated in at least one of the three sports. We add, $19 + 5 + 4 + 4 + 1 + 3 + 14 = 50$. Then, we subtract 50 from 70. Hence, 20 were students who participated in other sports.

As you can see, many more combinations can be considered. The Venn diagram is a useful aid in this type of survey, or counting, problem.

Exercises 1.5

1. Let A and B be sets such that $A \cap B = \phi$. Draw a Venn diagram to represent the following:
a) $A \cap B$
b) $A \cup B$

2. Let A and B be sets. Draw a Venn diagram for each of the following:

a) $A - B$ d) $\overline{\overline{A}} \cup \overline{B}$ g) $\overline{A \cap \overline{B}}$
b) $\overline{A} - \overline{B}$ e) $(A - B) - \overline{B}$ h) $\overline{A - B}$
c) $\overline{\overline{A} \cup B}$ f) $\overline{A \cap B}$ i) $A \cup \overline{B}$

3. Let R, S, and T be subsets of a universal set U. Draw a Venn diagram to represent each of the following sets.

a) $R \cap (S \cup T)$ f) $R \cap (S \cap T)$
b) $(R \cap S) \cup (R \cap T)$ g) $(R \cap S) - T$
c) $R \cup (S \cap T)$ h) $R - (S \cap T)$
d) $(R \cup S) \cap (R \cup T)$ i) $[(R \cup S) - T] \cap R$
e) $(R \cap S) \cap T$ j) $[(R \cup S) - (R \cap S)] \cup T$

4. A Ford dealer, a Chevrolet dealer, and a VW dealer

conducted a survey of 150 car owners in the town of Automobile. The results of the survey follow.

43 owned Fords
38 owned VWs
56 owned Chevrolets
18 owned Chevrolets and VWs
21 owned Fords and VWs
5 owned Fords and Chevrolets
5 owned all three

a) How many people surveyed owned only a Ford? a Chevrolet?
b) How many people owned a Ford and Chevrolet but not a VW?
c) How many did not own a Ford, Chevrolet, or VW?

5. A survey was made of 130 children to see what their reaction was to certain vegetables. The results were:

36 liked spinach
43 liked turnips
37 liked cabbage
30 liked turnips and cabbage
22 liked spinach and cabbage
19 liked spinach and turnips
18 liked all three vegetables

a) Construct a Venn diagram summarizing these data.
b) How many children liked only turnips?
c) How many children liked spinach but did not like turnips or cabbage?
d) How many children didn't like any of the three vegetables?

6. In gathering information for a curriculum meeting of an English department, it was necessary to take a survey to see what kind of literature the incoming freshmen had read. From the 200 students surveyed, the statistical results revealed the following:

143 students had read *The Scarlet Letter*
23 students had read *War and Peace*
64 students had read *Huckleberry Finn*
30 students had read *The Scarlet Letter* and *Huckleberry Finn*

21 students had read *War and Peace* and *Huckleberry Finn*
20 students had read all three books
20 students had read *The Scarlet Letter* and *War and Peace.*

a) Construct a Venn diagram summarizing these data.
b) How many students had not read any of the books?
c) How many students had read only one book? only two books?
d) How many students had read only *War and Peace*?

7. Construct a Venn diagram showing all 16 possible regions that can exist among 4 sets.

8. Using the Venn diagram constructed above, solve the following problem. A survey was taken of 103 students enrolled in a home economics class in interior decorating. The students were asked to rate the colors red, yellow, orange, and green according to personal like or dislike. Here are the results.

54 liked orange
41 liked red
63 liked yellow
37 liked green
25 liked red and yellow
18 liked red and green
23 liked yellow and green
28 liked yellow and orange
14 liked orange and green
10 liked red, yellow, and orange
12 liked red, yellow, and green
7 liked yellow, orange, and green
18 liked red and orange
8 liked red, orange, and green
3 liked red, yellow, orange, and green

a) How many students liked only yellow?
b) How many liked only red?
c) How many students didn't like any of the colors?
d) How many liked only yellow and orange?
e) How many liked only red, green, and orange?
f) How many students liked only two colors?

9. Construct a Venn diagram for A Δ B as defined in Exercises 1.4, Problem 24.

Section 6
Sameness

We now have much useful and interesting information regarding ways to express sets. A natural question to consider is, "How do we know when two different expressions denote equal sets?" For instance, when A, B, and C are subsets of U, the set $A \cup (B \cap C)$ is the same as the set $(A \cup B) \cap (A \cup C)$ even though the expressions are different.

Let's check to see if $A \cap (B \cap C) = (A \cap B) \cap C$ for $U = \{1,2,3,\ldots,9\}$, $A = \{1,2,3,4\}$, $B = \{1,3,4,5\}$, and $C = \{3,4,7,8\}$. First, consider $A \cap (B \cap C)$. Now, $B \cap C = \{3,4\}$ and $A \cap (B \cap C) = A \cap \{3,4\} = \{3,4\}$. Next, consider $(A \cap B) \cap C$, and note that $A \cap B = \{1,3,4\}$ and $(A \cap B) \cap C = \{1,3,4\} \cap C = \{3,4\}$. Hence, in this example

$$A \cap (B \cap C) = (A \cap B) \cap C$$

because the two sets have exactly the same elements.

If A, B, and C are subsets of a universal set U, such that A, B, C, and U are not specified, how can we be sure that the asserted equality, $A \cap (B \cap C) = (A \cap B) \cap C$, is true? There are several acceptable methods to show that sets are equal. The first method we will discuss is the *element inclusion* method, a method suggested by the definition of set equality. Recall that for sets A and B to be equal, $A \subset B$ and $B \subset A$. In practice this means we must pick some arbitrary element of A (call it x) and show that this x is also an element of B. Then, we must turn things around and show that some arbitrary element of B, say z, is also an element of A.

Let's apply the element inclusion method to the asserted equality $A \cap (B \cap C) = (A \cap B) \cap C$. We make no restrictions on the sets A, B, and C. All we know is that A, B, and C are sets, nothing more. We have two jobs to do:

a) We must show that $A \cap (B \cap C) \subset (A \cap B) \cap C$.

b) We must show that $(A \cap B) \cap C \subset A \cap (B \cap C)$.

Part (a): Suppose x is any element in $A \cap (B \cap C)$. That is, $x \in A \cap (B \cap C)$. This means that $x \in A$ and $x \in B \cap C$. But $x \in B \cap C$ means $x \in B$ and $x \in C$. So $x \in A$ and $x \in B$ and $x \in C$. Since x is an element of both A and B, it is an element of their intersection. Hence, $x \in A \cap B$. Thus, $x \in A \cap B$ and $x \in C$. So x is an element of the intersection of $A \cap B$ and C, and we have $x \in (A \cap B) \cap C$. Since we picked x arbitrarily from $A \cap (B \cap C)$ and showed that it was in $(A \cap B) \cap C$, we have shown that any element of $A \cap (B \cap C)$ must be an element of $(A \cap B) \cap C$. That is, $A \cap (B \cap C) \subset (A \cap B) \cap C$.

Part (b): Suppose we pick an arbitrary $z \in (A \cap B) \cap C$. This means by the definition of intersection that $z \in A \cap B$ and $z \in C$. But $z \in A \cap B$ means $z \in A$ and $z \in B$. So we have $z \in A$ and $z \in B$ and $z \in C$. Now $z \in B$ and $z \in C$ means that z is an element of the intersection of B and C. So we have $z \in B \cap C$. However, $z \in A$ and $z \in B \cap C$ means that z must be in the intersection of A and $B \cap C$. Thus, $z \in A \cap (B \cap C)$. Hence, $(A \cap B) \cap C \subset A \cap (B \cap C)$.

Now by using parts (a) and (b), we satisfy the criterion specified in the definition for sets being equal so that $A \cap (B \cap C) = (A \cap B) \cap C$ (the *associative* property of intersection).

A second method for showing that sets are equal is by use of Venn diagrams. Let's apply this method to show $A \cap (B \cap C) = (A \cap B) \cap C$. Let A, B, and C be sub-

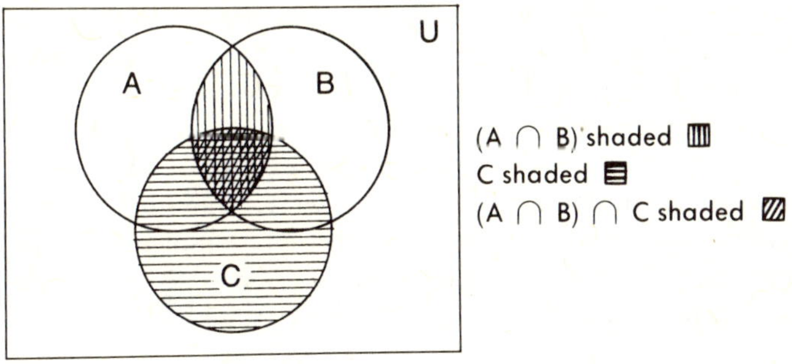

Figure 1.26

sets of U. The Venn diagram of $(A \cap B) \cap C$ is shown in Figure 1.26.

Figure 1.27 shows $A \cap (B \cap C)$.

Note that in Figures 1.26 and 1.27 the same regions are shaded in black. That is, $(A \cap B) \cap C$ has the same Venn diagram representation as $A \cap (B \cap C)$. Thus, we see that Venn diagrams provide us with a means of showing that two sets are equal. For many set-theoretic identities, the use of Venn diagrams provides an easy way to illustrate the plausibility of the identity. In general in mathematics, a diagram is not sufficient to establish the validity of a particular statement. However, the results of mathematical logic

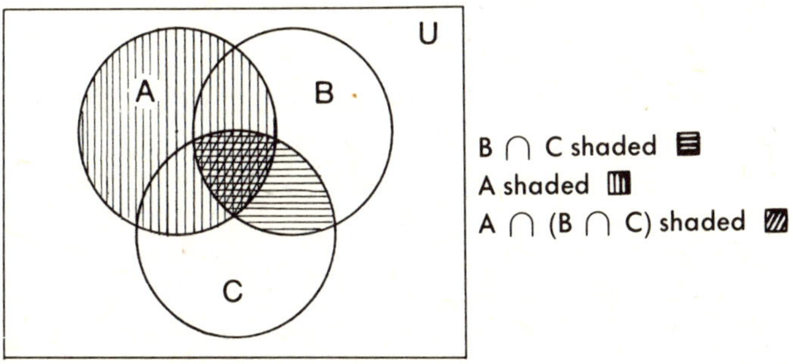

Figure 1.27

make it permissible to establish set-theoretic identities via Venn diagrams when Venn diagrams can adequately represent the sets involved in the identities.

There are several set-theoretic identities that occur frequently in mathematics. The identities listed below follow from the definition of the universal set U, the empty set ϕ, and the operations of union, intersection, complementation, and equality. Let U be a universal set and A, B, and C be subsets of U. Then the following identities hold.

Associative Laws

a) $A \cap (B \cap C) = (A \cap B) \cap C$
b) $A \cup (B \cup C) = (A \cup B) \cup C$

Commutative Laws

a) $A \cap B = B \cap A$
b) $A \cup B = B \cup A$

Distributive Laws

a) $A \cap (B \cup C) = (A \cap B) \cup (A \cap C)$
b) $A \cup (B \cap C) = (A \cup B) \cap (A \cup C)$

Complement Laws

a) $\overline{\overline{A}} = A$
b) $\overline{U} = \phi$
c) $\overline{\phi} = U$
d) $A \cup \overline{A} = U$
e) $A \cap \overline{A} = \phi$

Identity Laws

a) $A \cap \phi = \phi$
b) $A \cap U = A$
c) $A \cup U = U$
d) $A \cup \phi = A$

Exercises 1.6

1. Let A be a subset of the universal set U. Complete the following by using one of the symbols A, $\overline{A}$, U, or ϕ.

a) $A \cap U = ?$
b) $A \cap \phi = ?$
c) $A \cup \phi = ?$
d) $A \cup U = ?$
e) $A \cap A = ?$
f) $A \cup A = ?$
g) $A \cap \overline{A} = ?$
h) $A \cup \overline{A} = ?$
i) $U - A = ?$
j) $A - U = ?$
k) $A - A = ?$
l) $\phi - A = ?$
m) $A - \phi = ?$
n) $\overline{A - A} = ?$
o) $\overline{U} = ?$
p) $\overline{\phi} = ?$
q) $\overline{\overline{\phi}} = ?$
r) $\overline{\overline{U}} = ?$

2. Using Venn diagrams, show that the following are set identities.

a) $A \cup B = B \cup A$
b) $A \cap B = B \cap A$
c) $A \cup (B \cap C) = (A \cup B) \cap (A \cup C)$
d) $A \cap (B \cup C) = (A \cap B) \cup (A \cap C)$
e) $A \cup A = A$
f) $A \cap A = A$
g) $A \cap (A \cup B) = A$
h) $A \cup (A \cap B) = A$
i) $A - (B \cup C) = (A - B) \cap (A - C)$
j) $A - (B - C) = (A - B) \cup (A \cap C)$
k) $A \cap (B - C) = (A \cap B) - (A \cap C)$
l) $(A - B) - C = A - (B \cup C)$
m) $\overline{\bar{A} \cap (B \cap C)} = A \cup (\bar{B} \cup \bar{C})$

3. Use the distributive and commutative laws to find a set equal to each of the following.

a) $S \cap (R \cup T)$
b) $(R \cup S) \cap (R \cup T)$
c) $(S \cup T) \cap (R \cup T)$
d) $(A \cup B) \cap C$
e) $(A \cap B) \cup (A \cap C)$
f) $[(A \cup C) \cup A] \cap [(B \cup C) \cup A]$
g) $R \cup [(S \cap T) \cap (T \cap R)]$

4. Simplify the following set expressions by using identities.

a) $(\bar{A} \cap A) \cup (B \cap U)$
b) $(A \cup B) \cup (\bar{B} \cup \bar{\bar{A}})$
c) $[(A \cup \phi) \cup B] \cap \bar{B}$
d) $[(A \cup B) \cap B] \cap (\overline{A \cap B})$
e) $(B - A) \cap (\bar{A} \cup B)$
f) $[(A \cup B) \cap (A \cap B)] \cap (\overline{A \cap B})$

Example: Simplify $[A \cup (B \cap \bar{A})] \cup B$
$[A \cup (B \cap \bar{A})] \cup B = [(A \cup B) \cap (A \cup \bar{A})] \cup B = [(A \cup B) \cap U] \cup B = (A \cup B) \cup B = A \cup (B \cup B) = A \cup B$

5. The following identities are called De Morgan's Laws.

$$(1)\quad \overline{A \cup B} = \bar{A} \cap \bar{B}$$

$$(2)\quad \overline{A \cap B} = \bar{A} \cup \bar{B}$$

a) Verify these identities by using Venn diagrams.
b) Verify these identities by using the element inclusion method.
c) Can you make a conjecture as to De Morgan's Laws for three sets? That is,

$$\overline{(A \cap B) \cap C} = ? \quad \text{and} \quad \overline{(A \cup B) \cup C} = ?$$

6. Consider the sets A, B, C, D, and E, and the union or intersection of all of these sets. That is, consider

$$A \cup B \cup C \cup D \cup E \quad \text{or} \quad A \cap B \cap C \cap D \cap E.$$

Since union and intersection are binary operations, we must perform the union or intersection of these sets two sets at a time. Extended use of the associative laws reveals that it doesn't matter what pattern of grouping we choose. For instance,

$$\{[(A \cup B) \cup C] \cup D\} \cup E$$
$$= (A \cup B) \cup [(C \cup D) \cup E]$$

and

$$[A \cap (B \cap C)] \cap (D \cap E)$$
$$= [(A \cap B) \cap (C \cap D)] \cap E.$$

For example, let $A = \{1,2,3\}$, $B = \{3,4\}$, $C = \{1,3,5\}$, $D = \{2,3,5,6\}$, $E = \{1,3,6,8\}$. Then,

$$\{[(A \cup B) \cup C] \cup D\} \cup E$$
$$= \{[\{1,2,3,4\} \cup C] \cup D\} \cup E$$
$$= \{\{1,2,3,4,5\} \cup D\} \cup E$$
$$= \{1,2,3,4,5,6\} \cup E$$
$$= \{1,2,3,4,5,6,8\}$$

and,

$$(A \cup B) \cup [(C \cup D) \cup E]$$
$$= \{1,2,3,4\} \cup [\{1,2,3,5,6\} \cup E]$$
$$= \{1,2,3,4\} \cup \{1,2,3,5,6,8\}$$
$$= \{1,2,3,4,5,6,8\}$$

Hence,

$$\{[(A \cup B) \cup C] \cup D\} \cup E = (A \cup B) \cup [(C \cup D) \cup E].$$

In a similar manner, illustrate

$$[A \cap (B \cap C)] \cap (D \cap E) = [(A \cap B) \cap (C \cap D)] \cap E.$$

7. Let A, B, C, D, E, F, G be sets. Consider

$$A \cap B \cap C \cap D \cap E \cap F \cap G.$$

List seven different ways in which these sets can be grouped. Example:

$$[(A \cap B) \cap (C \cap D)] \cap [(E \cap F) \cap G]$$

8. Let $A = \{a,b,c,d\}$, $B = \{a,b,1,2\}$, $C = \{b,d,e\}$, $D = \{1,2,d,f\}$, $E = \{2,e,f\}$.

Using the above sets, illustrate the following.

a) $[(A \cup B) \cup (C \cup D)] \cup E = \{A \cup [B \cup (C \cup D)]\} \cup E$

b) $A \cup \{[(B \cup C) \cup D] \cup E\} = (A \cup B) \cup [C \cup (D \cup E)]$

c) $(A \cap B) \cap [(C \cap D) \cap E] = \{[(A \cap B) \cap C] \cap D\} \cap E$

d) $[(A \cap B) \cap C] \cap (D \cap E) = A \cap [(B \cap C) \cap (D \cap E)]$

9. Using the element inclusion method, show that:

a) $A \cup B = B \cup A$.

b) $A - (B \cup C) = (A - B) \cap (A - C)$.

c) $A \cap (B \cup C) = (A \cap B) \cup (A \cap C)$.

d) $A - (B \cap C) = (A - B) \cup (A - C)$.

e) $A \cap B = B \cap A$.

f) $A \cup (A \cap B) = A$.

g) $A \cup (B - A) = A \cup B$.

Section 7
Beyond Large

If you were asked to define an infinite set, you would probably say something like, "An infinite set is a set that has an unlimited number of elements in it." This intuitive notion is good enough for everyday purposes, but in mathematics we make the idea more precise. The following discussion will help make our definition of infinite and finite sets plausible.

Consider the sets $N = \{1, 2, 3, \ldots\}$ and $N' = \{2, 3, 4, \ldots\}$. Match their elements as indicated in Figure 1.28. Notice that N' is a proper subset of N and that we can indicate a one-to-one correspondence between N and N'. (See if you can find other matchings with N and a proper subset of N.) The fact that a set can be equivalent to a proper subset of itself was taken by Georg Cantor* to be the defining characteristic of an infinite set.

Definition: A set is an *infinite* set if and only if it is equivalent to a proper subset of itself.

Example 8: Is the set $N_{even} = \{2, 4, 6, 8, \ldots\}$ an infinite set? Consider the set $N_{four} = \{4, 8, 12, 16, \ldots\}$. Notice that $N_{four} \subset N_{even}$ and $N_{four} \neq N_{even}$. A one-to-one correspondence is illustrated as follows.

*Georg Cantor (1845–1918) published his first revolutionary paper on the theory of infinite sets at the age of 29.

$$N_{even} = \{2, 4, 6, 8, \ldots\}$$
$$\;\;\;\;\;\;\;\;\;\;\;\;\;| \;\; | \;\; | \;\; |$$
$$N_{four} = \{4, 8, 12, 16, \ldots\}$$

$$N = \{1, 2, 3, \ldots, n, \ldots\}$$
$$\;\;\;\;\;\;\;| \;\; | \;\; | \;\;\;\;\;\;\; |$$
$$N' = \{2, 3, 4, \ldots, n+1, \ldots\}$$

Figure 1.28

Hence, N_{even} is an infinite set.

Let A = {a, b, c}. Try to set up a one-to-one correspondence between A and a proper subset of itself, say {a, b}. The attempt to do so is futile. This illustrates the fact that a finite set cannot be put into a one-to-one correspondence with a proper subset of itself. We make our definition of finite sets in terms of infinite sets.

Definition: A set is finite if and only if it is not infinite.

We mention here that any two infinite sets are not necessarily equivalent. Cantor showed that N = {1, 2, 3, . . .} is not equivalent to the set of points on a line. To do this, he showed the impossibility of finding a one-to-one correspondence between these two sets by pointing out that no matter what matching is used, one can always find some point on the line that is not matched with any one of the numbers in N. (For further discussion of this topic, check the references at the end of the chapter.)

We wish to standardize the notation we will use when referring to some important infinite sets. The notation is as follows.

N = natural numbers or counting numbers = {1, 2, 3, . . .}

W = whole numbers = {0, 1, 2, 3, . . .}

Z = integers = {. . . , −3, −2, −1, 0, 1, 2, 3, . . .}

$$Q = \text{rational numbers} = \left\{ \frac{x}{y} \mid x \in Z \text{ and } y \in N \right\}$$

R = real numbers = $\{x \mid x$ is any number that can be associated with a point on the number line$\}$

A number line is a line on which one marks off equally spaced points and labels them as illustrated below.

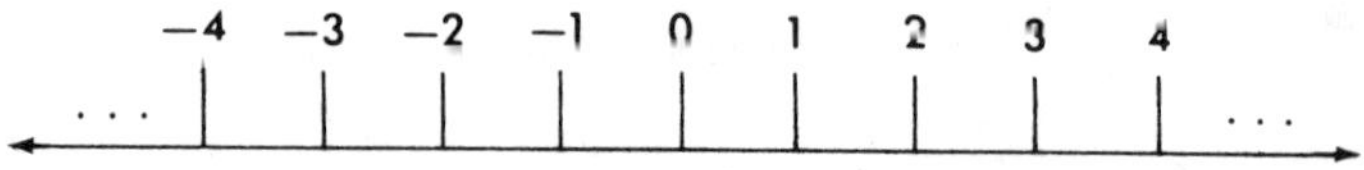

Exercise 1.7

1. Let I and F stand for infinite and finite, respectively. Indicate whether the sets that follow are infinite or finite.
a) The set of keys on a typewriter
b) The set of grains of sand along the Pacific coast
c) The set of lines that can be drawn through a specific point
d) The set of hairs on your head
e) The set of all multiples of 3
f) The set of all triangles that can be inscribed in a given circle
g) The set of all people with wings
h) The power set of all points in a line
i) The set of points in a circle
j) The set of all cactus growing in Arizona

2. True or False.
a) Given two finite sets, A and B, then it is possible that $A \cap B$ is infinite.
b) Given a finite set A and an infinite set B, then it is necessary that $A \cup B$ is infinite.
c) It is possible to have two finite sets, A and B, such that $A \cup B$ is infinite.
d) Let A be a finite set and let B be an infinite set. Then, $A \cap B$ must be infinite.
e) Let A be a finite set and let B be an infinite set. Then $A - B$ must be a finite set.
f) Let A be a finite set and let B be an infinite set. Then $B - A$ must be infinite.
g) Let A and B be two infinite sets. Then, $A \cup B$ must be infinite.

h) Let A and B be two infinite sets. Then, $A \cap B$ must be infinite.

i) Let A and B be two infinite sets. Then $B - A$ must be finite.

3. Indicate whether each of the following is true or false.

a) $N \subset W$

b) $W \subset Z$

c) $N \subset Q$

d) $W - N = \phi$

e) $\mathbf{R} = Q \cup \mathbf{R}$

f) $Q \cap \mathbf{R} = \{0\}$

g) $W \cup Z = W$

h) $Q \subset \mathbf{R}$

Section 8
Figments of the Mind

The word *number* is generally used in an undefined sense. You have relied on your intuition about what a number is, and your intuition has probably served you well. However, suppose you want to tell someone precisely what you mean by *number*. Upon reflection, you might find it difficult to pin down a precise definition.

Is the symbol "2" the number two? Are any of the following symbols the number two: II, $4\int_0^1 x\,dx$, 4-2, $8 \div 4$? None of these symbols is the number two. They are only representations of the number two. We call these representations *numerals,* not numbers. Numbers are concepts; that is, they are abstract entities that are objects of our thoughts. What we intend to do now is to consider the nature of the numbers themselves, the concepts that are represented by numerals.

How does a child learn the meaning of "hot"? As the child grows up, he hears the word "hot" associated with different objects. That is, when he tries to touch a stove, he is told, "Hot!" When he approaches a furnace, he is told, "Hot!" If he gets close to a fire, he is told "Hot!" He soon learns that many things may be hot—stoves, furnaces, fires, etc. All of these objects have the property of hotness, yet not any one of them is the concept "hot." In a similar manner, a child learns the property of twoness by being shown two fingers, two toys, two cars, etc.

Let A = {a,b}, B = {John,Joe}, C = {thumb,index finger} and D = {$, %}. We can establish a one-to-one correspondence between any two of these sets, that is, each of sets A, B, C, and D is equivalent to any of the other sets. The equivalence of the sets A, B, C, and D demonstrates a common property shared by each set, namely, that of having two elements.

The following definition of cardinal number embodies the ideas of the previous discussion and is a reflection of the manner in which one forms the concept of number.

Definition: The *cardinal number* of a set is that property of the set that it has in common with all sets that are equivalent to it. For a set A we denote the cardinal number of A by the symbol $n(A)$.

For finite sets, we adopt the usual symbols for the whole numbers as the symbols for the cardinal numbers of these sets. For example, if A = {r,s,t}, then $n(A) = 3$. We give the name zero to the cardinal number for the empty set. The cardinal number of a finite set tells us how many objects are in that set.

A notion that is closely associated with that of cardinal number is that of counting. Counting is an integral part of our daily activities. Exactly what are we doing when we count? To aid us in the discussion of counting, we will construct the following sets. Let A be the set that has a as its element, A = {a}, let B = {a,b} and let C = {a,b,c}. Continuing in this manner, we can form larger sets by adding one element at a time. The symbols for the cardinal numbers associated with these sets are $n(A) = 1$, $n(B) = 2$, $n(C) = 3$, and so on. Notice that

$$\{a\} \subset \{a,b\} \subset \{a,b,c\} \subset \ldots .$$

We can continue this series indefinitely. (After we run out of letters, we can use other objects.) Recall that $n(\phi)$ is zero, and note that

$$\phi \subset \{a\} \subset \{a,b\} \subset \{a,b,c\} \subset \dots.$$

Now arrange the names of the cardinal numbers assigned to the above sets in order from left to right as determined by the position their respective sets occupy in the sequence of subsets, $\phi \subset A \subset B \subset C \subset \dots$. In this way, we obtain the infinite *ordered* set

$$\{0,1,2,3,\dots\},$$

which is the set of whole numbers. From this set of cardinal numbers, we form the ordered subset of natural numbers or counting numbers, $N = \{1,2,3,\dots\}$. The following *ordered* subsets of N are called *counting sets*.

$$\{1\}, \{1,2\}, \{1,2,3\}, \dots, \{1,2,3,\dots,n\}, \dots$$

Now, we return to our question about counting. Consider the set $C = \{a,b,c\}$. We "count" this set by looking at a and thinking one. Similarly, we look at b and think two. Then, we look at c and think three. It is clear that we have matched the elements of the given set and the elements of the counting set $\{1,2,3\}$, and so we say that the given set contains three objects, three being the last cardinal number in that counting set. The preceding discussion should make the following definition of counting reasonable.

Definition: ***Counting*** **is a process that sets up a one-to-one correspondence between a set of objects and the elements of a finite counting set $\{1,2,3,\dots,n\}$.**

For example, suppose we want to count the letters in the word *gone*. We select a counting set that establishes a one-to-one correspondence with the letters in the word *gone:*

$$\begin{matrix} \{g, & o, & n, & e\} \\ | & | & | & | \\ \{1, & 2, & 3, & 4\} \end{matrix}$$

The cardinal number we assign to the set $\{g,o,n,e\}$ is four, the last cardinal number in that counting set.

In the definition of cardinal number, we did not require that the set be finite. Every set can have a cardinal number, whether the set be finite or infinite. The only requirement for two sets to have the same cardinal number is that they be equivalent. The cardinal numbers for infinite sets are called *transfinite cardinal numbers*.

Cantor called the cardinal number of the set

$$\{0, 1, 2, 3, 4, \ldots\}$$

"aleph-sub-nought," written $\aleph_0$. Thus, $n(\{0, 1, 2, 3, 4, \ldots\}) = \aleph_0$. As we mentioned earlier, we cannot necessarily establish a one-to-one correspondence between any two infinite sets. The points on a line have a different transfinite cardinal number than the set of whole numbers. The transfinite cardinal number of the set of points on a line is usually denoted by "c," as is the transfinite cardinal number of the set of real numbers. This c is a larger transfinite cardinal number than $\aleph_0$.

It has been shown that the cardinal number of a set $\mathcal{P}(A)$ is always larger than the cardinal number of the set A. Hence, by putting into action our previous notion of power sets, we can come up with a scheme for still larger transfinite cardinal numbers. Let S be an infinite set. Then $\mathcal{P}(S)$ is an infinite set with a larger transfinite cardinal number than S. Continuing with $\mathcal{P}(\mathcal{P}(S))$, $\mathcal{P}(\mathcal{P}(\mathcal{P}(S)))$, etc., we see that each of these sets has a cardinal number greater than the one before it. From this pattern we see that there are infinitely many transfinite cardinal numbers. Of course, we have not "proved" any of the above statements because to show these concepts about transfinite cardinal numbers rigorously is beyond what we can do in this book.

Perhaps all this discussion on transfinite cardinal numbers leaves you muttering to yourself. If so, do not despair. We only mention these ideas to expand your concept of cardinal number and illustrate how the notion of sets makes such reasoning possible.

As you are aware, we use whole numbers in other ways besides just to give a "count" of a set or to indicate "size." We also use whole numbers to indicate rank or position. An *ordinal number* indicates the rank or position of an object according to some pattern. For example, suppose you are telling someone about a friend's graduation. You might say,

CARDINAL AND ORDINAL NUMBERS	
Cardinal	**Ordinal**
I have 2 eyes. January has 31 days. She called 5 times. I read 120 pages. We bought 8 chickens.	Read page 81 in the book. This is the 5th time. June 19. He died in 1966. 3205 N. 22nd Avenue.

Figure 1.29

"Jim was nineteenth in his graduating class." In this case, you are using nineteen to indicate the rank of Jim as compared to his classmates. Here nineteen is used in the ordinal sense. Basically, a cardinal number answers the question "How many?" and an ordinal number answers the question "In what position?" For example, if you say, "Look on the 175th page of the book," you name an ordinal number. However, if you say, "The book has 176 pages," you name a cardinal number.

Exercises 1.8

1. Give the cardinal number for each of the following sets.
a) $A = \{1, 2, 3, \ldots, 10\}$
b) $A = \{x \mid x$ is a letter of the English alphabet$\}$
c) $A = \{y \mid y$ is a key on the piano$\}$
d) $A = \{y \mid y$ is a letter in the word *fruitcake*$\}$
e) $A = \{2, 4, 6, \ldots, 100\}$
f) $A = \{3, 6, 9, \ldots, 999\}$

2. For each of the following, state whether a cardinal (C) or an ordinal (O) number has been named.
a) Send 130 copies.
b) February 14
c) Read 18 pages.
d) 1776 was an important year.

e) We milked 8 cows.
f) I was 12 when it happened.
g) 18 days ago
h) Page 36 has the correct drawing.
i) The 3rd try was the charm.
j) I shot par on hole No. 18.

3. Let C be the set of points on the circle in the figure and let T be the set of points on the triangle. Show that C ≡ T by displaying a one-to-one correspondence between C and T. What is the relationship between n(C) and n(T)?

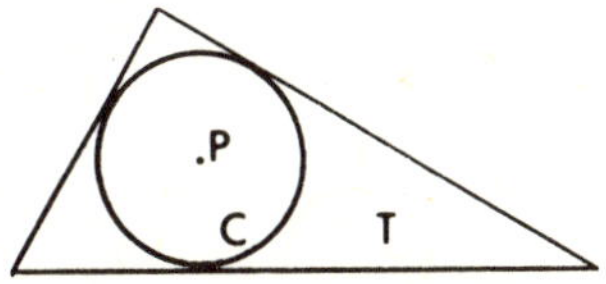

4. Let C and C′ be two concentric circles; that is, two circles that have the same center. Let A be the set of points in circle C and let B be the set of points on circle C′. Display a one-to-one correspondence between A and B. What is the relationship between n(A) and n(B)?

5. Let A = {1,2,3,...}, and let B = {5,10,15,...}.
a) Set up a one-to-one correspondence between sets A and B.
b) Under your correspondence, what element in B would you pair with 199 ∈ A?
c) What is n(B)?

6. Let A = {1,11,111,1111,...}, and let B = {10,100, 1000, 10000,...}.
a) Set up a one-to-one correspondence between A and B.
b) Consider 1000000 ∈ B. According to your correspondence, with what element of A would you pair 1000000?
c) What is n(A)?
d) What is n(B)?

7. Let A = {x, y, z}.
a) What is n(A)?
b) What is $n(\mathcal{P}(A))$?
c) What is $n(\mathcal{P}(\mathcal{P}(A)))$?
d) What is $n(\mathcal{P}(\mathcal{P}(\mathcal{P}(A))))$?

Suppose A is a set with k elements; that is, let k designate the number of elements in a set A. Then, $n(A) = k$. Can you make a conjecture as to the number of elements in $\mathcal{P}(\mathcal{P}(A))$? Give a general formula.

8. Is there a largest cardinal number? If so, what is it?

Section 9
Familiar Concepts in New Clothes

Addition

Two children decide to go to the store to buy candy. Being aware that they need a certain number of pennies to get the candy they want, they decide to count the pennies they have. Child A has six pennies. Child B has eight pennies. How do they determine the number of pennies they have altogether? They probably put both sets of pennies together and count the pennies in the new set that consists of child A's pennies and child B's pennies. If we put this illustration into our set notation and terminology, we have the following.

Child A has a set of six pennies. Call the set A, and let

$$A = \{P_1, P_2, P_3, P_4, P_5, P_6\}.$$

We note that $n(A) = 6$. Similarly, for child B we have

$$B = \{P_1', P_2', P_3', \ldots, P_8'\}$$

and $n(B) = 8$. Now,

$$A \cup B = \{P_1, P_2, \ldots, P_6, P_1', P_2', \ldots, P_8'\},$$

and $n(A \cup B) = 14$. Now we can write for this example,

$$n(A) + n(B) = n(A \cup B), \quad \text{or} \quad 6 + 8 = 14.$$

We have found a way of stating the sum of two cardinal numbers in terms of our set-theoretic concepts.

The next question is, "For sets A and B, will $n(A) + n(B) = n(A \cup B)$ always be true?" Further investigation will reveal that we have to make sure sets A and B are disjoint. For example, consider the sets $A = \{1,2,3,4\}$ and $B = \{3,4,5,6\}$. Then $A \cup B = \{1,2,3,4,5,6\}$. We have $n(A) = 4$, $n(B) = 4$, and $n(A \cup B) = 6$. But we know that $4 + 4 \neq 6$. Hence, if we want to write a definition of the sum of two cardinal numbers, we had better be sure to include the condition of disjointness.

In our definition of the sum of two cardinal numbers, we will make some subtle assumptions. We will assume that if a and b are any two cardinal numbers, then there exist disjoint finite sets A and B such that $n(A) = a$ and $n(B) = b$. With these preliminary remarks, we are ready to make our definition.

Definition: The *sum* of the cardinal numbers a and b, written $a + b$, is the cardinal number $n(A \cup B)$ where $n(A) = a$, $n(B) = b$, and $A \cap B = \phi$. Symbolically, we write

$$a + b = n(A) + n(B) = n(A \cup B).$$

Example 9: Show that $2 + 3$ is equal to the cardinal number 5. Let $A = \{a,b\}$ and $B = \{@,*,\#\}$. Then $A \cap B = \phi$ and $A \cup B = \{a,b,@,*,\#\}$. We have $n(A) = 2$, $n(B) = 3$, and $n(A \cup B) = 5$, so that $2 + 3 = n(A) + n(B) = n(A \cup B) = 5$.

Notice that our definition does not depend on the disjoint sets we pick. In the preceding example, we could have chosen $A = \{\text{Joe}, \text{Jack}\}$ and $B = \{x,y,z\}$, and we could still have shown that $2 + 3$ equals the cardinal number 5. Since our definition is independent of the particular disjoint sets A and B that we choose as representations, we are assured that $a + b$ is well defined. That is, for any two cardinal numbers a and b, $a + b$ will be a unique cardinal number.

Multiplication

Can you think of a way to define multiplication of cardinal numbers with our present set operations? You could do this, but it would be rather cumbersome. To facilitate a definition of multiplication, which allows multiplication to be completely independent from addition and which is in terms of an operation on sets, we need to develop a new method of assigning another set to a pair of sets.

So far, we have considered only sets whose elements are abstract ideas, persons, numbers, letters, etc. We now want to consider sets whose elements are ordered pairs.

Suppose you have bought a new car. You are taking it on a trip and want to check the gas mileage each time you get gas. You probably would make a chart of figures similar to the chart in Figure 1.30.

Now, each distance in miles is paired with gallons of gas consumed. You might write this information compactly as (170, 8), (210, 10), (265, 13), and (154, 7). In each case you put the miles traveled first and gallons of gas consumed second. It is important that you keep these numbers paired in the stated order to have meaningful information. If you figure gas mileage for 8 miles distance and 170 gallons of gas consumed (8, 170), you would probably decide that your car needed a tune-up!

In general, by an *ordered pair* of objects we mean a pair in which one of the objects is agreed upon to be the first object and the other to be the second. If a and b designate the objects, then (a, b) denotes the pairing of a and b where a denotes the object that must come first and b denotes the other object. Notice that $(a, b) \neq \{a, b\}$; (a, b) is an ordered pair, while $\{a, b\}$ is a set whose elements are a and b. In general, (a, b) is different from (b, a). In fact, $(a, b) = (x, y)$ if and only if $a = x$ and $b = y$. With this concept of an ordered pair

Miles Traveled	*Gallons of Gas Used*
170	8
210	10
265	13
154	7

Figure 1.30

in mind, we are ready to define a new operation on sets called Cartesian product.*

Definition: If A and B are sets, then the set of all ordered pairs (a,b), such that $a \in A$ and $b \in B$, is called the *Cartesian product* (or cross product) of A and B, denoted $A \times B$ and read "A cross B". That is,

$$A \times B = \{(a,b) \mid a \in A \text{ and } b \in B\}.$$

In the ordered pair (a,b), a is called the *first coordinate* and b is called the *second coordinate*.

Example 10: Let $A = \{a,b,c\}$ and $B = \{1,2\}$. Then, $A \times B = \{(a,1), (a,2), (b,1), (b,2), (c,1), (c,2)\}$, and

$$B \times A = \{(1,a), (1,b), (1,c), (2,a), (2,b), (2,c)\}.$$

Notice that $A \times B \neq B \times A$. The sets do not have exactly the same ordered pairs, although they do have the same number of ordered pairs.

A useful technique for actually computing the Cartesian product of two specific sets, A and B, such as in the above example, is the following. To get $A \times B$, take the first letter of set A and pair it with each letter of B. Let the elements of B be the second coordinate. That is,

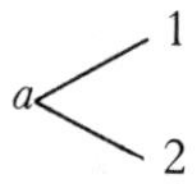

yields (a, 1) and (a, 2)

Similarly, for the second and third elements of A,

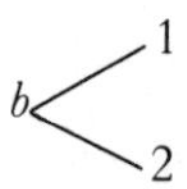

yields (b, 1) and (b, 2)

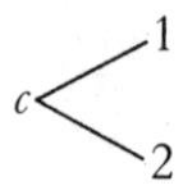

yields (c, 1) and (c, 2)

Then, collecting all of these ordered pairs, we have

*The Cartesian product is named after René Descartes, (1596–1650), a French philosopher and mathematician.

$$A \times B = \{(a,1), (a,2), (b,1), (b,2), (c,1), (c,2)\}$$

Perhaps you see the connection between the Cartesian product and a definition for multiplication of cardinal numbers. Using the sets in our above example, we illustrate this connection. We know that $n(A) = 3$, $n(B) = 2$, and, since $A \times B$ is a set having six ordered pairs as its elements, $n(A \times B) = 6$. The product of the cardinal numbers 3 and 2, written $3 \cdot 2$, is found by forming the Cartesian product from sets A and B and then noting the cardinal number of $A \times B$.

Definition: The *product* of the cardinal numbers a and b, written $a \cdot b$, is the cardinal number $n(A \times B)$ where $n(A) = a$ and $n(B) = b$. Symbolically, we write $a \cdot b = n(A) \cdot n(B) = n(A \times B)$

Example 11: Find the product of 3 and 4. First, we choose a set A that has cardinal number 3. Say $A = \{a,b,c\}$. Then we choose a set B that has cardinal number 4. Say $B = \{1,2,3,4\}$. We then form the Cartesian product of A with B:

$$A \times B = \{(a,1), (a,2), (a,3), (a,4), (b,1), (b,2), (b,3), (b,4), (c,1), (c,2), (c,3), (c,4)\}$$

Since $n(A \times B) = 12$, we have $3 \cdot 4 = n(A \times B) = 12$.

Exercises 1.9

1. Let $A = \{p,r,s\}$ and $B = \{5,6,7\}$. Write the following:

a) $A \times B$ c) $B \times A$
b) $A \times A$ d) $B \times B$

2. a) Let $A = \{\text{tree}\}$ and $B = \{\text{red, blue, white, green}\}$. Using sets A and B and the set-theoretic definition of addition, show that $1 + 4 = 5$.
b) Let $A = \{\$\}$ and $B = \{21,32,43,54\}$. Using these sets and the set-theoretic definition of addition, show that $1 + 4 = 5$.
c) What property of addition of cardinal numbers is illustrated by (a) and (b)?

3. Using the set-theoretic definition of addition, show the following.

a) $2 + 2 = 4$
b) $3 + 5 = 8$
c) $4 + 0 = 4$
d) $6 + 7 = 13$
e) $20 + 13 = 33$
f) $1 + 1 = 2$
g) $0 + 1 = 1$
h) $0 + 0 = 0$

4. Using the set-theoretic definition of multiplication, show that

a) $1 \cdot 0 = 0$
b) $2 \cdot 2 = 4$
c) $1 \cdot 1 = 1$
d) $4 \cdot 2 = 8$
e) $3 \cdot 1 = 3$
f) $0 \cdot 0 = 0$
g) $4 \cdot 3 = 12$
h) $2 \cdot 5 = 10$

5. If $A \times B = B \times A$, what do you know about sets A and B?

6. If $A \times B = \phi$ and $B \neq \phi$, what do you know about set A?

7. Let $U = \{a, e, i, o, u, 2, 4\}$, $A = \{2, 4\}$, $B = \{a, e, i\}$, $C = \{o, u\}$. Using these sets, show the following.

a) $A \times (B \cup C) = (A \times B) \cup (A \times C)$
b) $A \times (B \cap C) = (A \times B) \cap (A \times C)$
c) $A \times B = (A \times U) \cap (U \times B)$
d) $A \times (B - C) = (A \times B) \cap (A \times \overline{C})$

8. Let $A = \{1, 2\}$, $B = \{a, b, c\}$, $C = \{a, b\}$, and $D = \{¢, \$\}$.

a) What is $n(A)$? $n(C)$? $n(D)$?
b) What is $n(B)$?
c) Are A, C, and D disjoint?
d) Show that $n[(A \cup C) \cup D] = n(A \times B)$.

Note that this shows that $(2 + 2) + 2 = 2 \cdot 3$

9. Let A, B, and C be subsets of a universal set U. Using the element inclusion method, show that

a) $A \times (B \cup C) = (A \times B) \cup (A \times C)$
b) $A \times (B - C) = (A \times B) \cap (A \times \overline{C})$

10. Let $A = \{1, 2, 3, \ldots\}$, and $B = \{@, \#, \$\}$.

a) What is $n(A)$?
b) What is $n(B)$?
c) Is $A \cap B = \phi$?
d) Set up a one-to-one correspondence between A and $A \cup B$.
e) What is $n(A \cup B)$?
f) What is $\aleph_0 + 3$?

Section 10
Sentimental Remarks

The concepts presented in this chapter are basic. They provide a springboard from which we can investigate many interesting mathematical structures. During the whole development, emphasis is placed on how set-theoretic concepts form a mathematical tapestry that provides continuity to the topics studied.

The fundamental nature of the theory of sets should be well illustrated by the application of set concepts to so many ideas that are quite basic in themselves. The notions of numbers, counting, addition, and multiplication have been cast in set-theoretic terms. The idea of an infinite collection of objects has been characterized in a concise way. We hope the student will gain an appreciation for the beauty of such a primitive notion as set as he sees more and more ideas woven together with set-theoretic concepts.

References
Chapter One

Armstrong, James W., *Elements of Mathematics* (London: The Macmillan Company, 1970), pp. 87–91.

Hildebrand, Francis H., and Johnson, Cheryl G., *Finite Mathematics* (Boston: Prindle, Weber & Schmidt, Inc., 1970), pp. 12–15.

Pinter, Charles C., *Set Theory* (Menlo Park, Calif.: Addison-Wesley Publishing Co. Inc., 1971), pp. 3–4.

Stein, Sherman K., *Mathematics: The Man-Made Universe* (San Francisco: W. H. Freeman and Company Publishers, 1963), pp. 252–256.

Stoll, Robert R., *Set Theory and Logic* (San Francisco: W. H. Freeman and Company Publishers, 1963), pp. 79–90.

Newman, James R., ed., *The World of Mathematics*, vol. IV (New York: Simon & Schuster, Inc., 1956), pp. 2315–2347.

CHAPTER TWO

Logic

Nearly every day we are confronted with problem situations. In resolving these daily problems, we are required to use our reasoning ability. Language is the vehicle through which we express thoughts; it is the raw material that we use in the reasoning process. The more precise we are in our use of language, the better we can reason and express ourselves.

All of us can reason without the aid of specific training in logic. In the preceding chapter, we provided convincing arguments for the acceptance of several statements without knowing the formal techniques of logic. The basic tenets of reasoning are inherent within our language. The reasoning process, however, is often encumbered by vague and equivocal notions about the words used and the emotional impact of the style. Suppose we strip away the problems of meaning and interpretation. We are still faced with the necessity of determining whether or not we have reasoned correctly. This is the point at which symbolic logic enters the picture. Statements that occur in daily discourse must be translated into symbolic language free from encumbering ambiguities. Once this has been done, we can use the principles of symbolic logic to assess our reasoning.

The processes by which man thinks or reasons were first systematized by Aristotle (384–322 B.C.) in his work *Organon*. George Boole* is generally credited with being the father of modern logic. He expanded the development of logic through the use of algebraic operations.

Through the use of symbolic logic we can determine correct reasoning mechanically. We can almost say that by using symbolic logic we can reason correctly without thinking.

*George Boole (1815–1864) had little formal schooling. In 1854 he published *An Investigation of the Laws of Thought*, a revolutionary work on which the mathematical theories of logic and probability are founded.

Section 1
Sentences and Statements

To be subject to logical discourse, a sentence must be either true or false. We must further require that it be clear and precise enough in a given context to take on exactly one of these truth values. That is, it must be either true or false; it cannot be both true and false. This type of sentence will be given a special name.

Definition: A *statement* is any declarative sentence for which it is possible to consider the sentence to be either true or false in a given context.

If a statement is true, we say its *truth value* is true (written T), and if a statement is false, we say its truth value is false (written F).

Example 1:

Statement	*Truth Value*
a) Birds are animals.	T
b) Chicago has a population of exactly 100 people.	F
c) Some babies cry.	T

We will use capital letters, P, Q, R, etc., to denote statements. For instance, the statement "It is a cold day" can be symbolized by P. That is, in a particular discussion we can let P represent the statement "It is a cold day."

We make the following distinction between simple and compound statements. A simple statement is a sentence which is composed of only one statement. A compound statement is a sentence composed of two or more simple statements.

Example 2:

Simple Statements

a) The time is 12:00.
b) $2 + 2 = 4$
c) The grass is green.
d) The moon is made of cheese.

Compound Statements

a) The time is 12:00, and $2 + 2 = 4$.
b) The grass is green, and the moon is made of cheese.
c) If we win the ball game, then we are district champions.
d) John is a gentleman and a scholar, or we have misjudged him.

Exercises 2.1

1. Indicate whether the following sentences are statements or not.

a) The automobile is mine.
b) Can you turn off the machine?
c) Department stores close at 6:00 P.M.
d) What is that?
e) My mother-in-law is in town, and I am sick.
f) Let's eat out.
g) Oh!
h) Run fast, or you will be caught.
i) If practice will help, then I shall practice.
j) It is ten o'clock, and the temperature is 37° Fahrenheit outside.

k) You can see better if you turn on the light.
l) There is a television show that is enjoyable to me.
m) "Give me liberty, or give me death."

2. Are all statements sentences?

3. List six sentences that are not statements.

4. Consider the sentence, "It is now 3:00 P.M."

a) Is this a statement?
b) What is the truth value of this sentence at

1) 1:00 A.M.	4) 10:00 A.M.
2) 12:00 noon	5) 3:00 P.M.
3) 3:01 P.M.	6) 2:59 P.M.

5. Consider the sentence, "Jupiter is an older planet than Saturn."

a) Is this a statement?
b) What is the truth value of this sentence?

6. Indicate whether the following statements are simple (S) or compound (C). If the statement is compound, indicate its component statements.

a) Peace of mind will come only if you are successful in your work.
b) The bank makes money if and only if people deposit their money in it.
c) My shoes are too small, but my feet don't hurt.
d) Don bought a new pair of shoes.
e) Judy and Liz are good friends.
f) Since we needed a couch, we started asking all of our friends if they had an extra one.
g) I cannot imagine anyone having the audacity to ask a friend if they could borrow a couch.
h) Horses were the only means of transportation available.
i) My car functions well if it has gas and if I change the oil occasionally.
j) Paul Revere devised a plan to alert the people.
k) It is not the case that the Declaration of Independence was signed in 1540.

7. Consider the following statements.

1) Homemade preserves are better than preserves bought in a grocery store.

2) Homemade preserves are not better than preserves bought in a grocery store.

a) Suppose you gave statement 1 a truth value of T. Then what truth value would you give statement 2?

b) Suppose you gave statement 1 a truth value of F. Then what truth value would you give statement 2?

Section 2
Open Sentences

In daily discourse, we encounter sentences that are not statements. Some sentences of this type are:

a) How are you?
b) Help!
c) Mary, can't you be more careful?
d) Watch out for the sandtrap!
e) My goodness, John!
f) Stop!

None of the above sentences can be converted into statements. However, there are some sentences that are not statements that can be converted into statements.

Throughout mathematics many sentences are used that contain *variables*—placeholders for a particular person, object, number, etc. For example, the sentence, "x is a man over six feet tall," contains the variable x. We cannot determine the truth value of this sentence because of the presence of the variable x. Thus, this is a sentence that is not a statement. When the variable x is replaced with a specific man, we can check to see if the resulting sentence is true or false. Suppose we replace the x with Joe Jones. Now we are able to determine the truth or falsity of "Joe Jones is a man over six feet tall" by simply determining Joe Jones's height. The sentence is now a statement.

We notice that we can replace x with any man; that is,

the set of replacement possibilities for the variable x is the set of all men. We call a set of possible replacements for a sentence variable a *replacement set* or *universal set* for that sentence. A sentence containing a variable is called an *open sentence*. We formalize the notion in the following definition.

Definition: An *open sentence* is a sentence that contains one or more variables, is not a statement, and can be converted to a statement by replacing the variable with an element or elements from a replacement, or universal, set.

Example 3: Let U = {1,2,3}. Consider the open sentence "x + 2 = 3." Note the following true or false statements that result from replacing x with 1, 2, or 3.

Statement	*Truth Value*
1 + 2 = 3	T
2 + 2 = 3	F
3 + 2 = 3	F

We shall use P(x), Q(x), R(x), etc., to denote open sentences having the variable x. If we replace the variable x with a particular element from a universal set, say a, then we write P(a), Q(a), R(a), etc. For example, if U = {2,4,6} and if P(x) is "$x + 3 = 7$," then:

P(2) is 2 + 3 = 7, a false statement

P(4) is 4 + 3 = 7, a true statement

P(6) is 6 + 3 = 7, a false statement

There is another technique by which an open sentence can be converted into a statement. Consider again the open sentence, "x is a man over six feet tall," and the associated universal set consisting of all men. Think of the following assertion: "There exists an x in U, such that x is a man over six feet tall." Since we can assign a truth value to this declarative sentence, it is a statement. (The truth value is T.) In a similar way we can make the assertion: "For all x in U, x is a man over six feet tall." Since we can assign a truth value to this declarative sentence, it is a statement. (The truth value is F.)

We have converted the open sentence "x is a man over six feet tall" to a statement by specifying a universal set and by prefixing the expression with "There exists an x in U such that" or "For all x in U." The prefixes "There exists an x in U such that" and "For all x in U" indicate quantity. They answer the question, "How many objects have the property described in the open sentence?" Since both expressions convey the idea of quantity, the words *all* and *there exists* (or words equivalent to *all* and *there exists*) are called quantifiers.

We call the quantifier "all" the *universal quantifier* and denote it by the symbol $\forall$. The quantification symbol $\forall$ is used before a variable, as $\forall\ x$. Let U be a universal set. The expression $\forall\ x \in \mathrm{U}$ represents several equivalent phrases. Some of the more common phrases are:

"For all x elements of U"

"For each x element of U"

"For every x element of U"

"For any x element of U"

We symbolize the universal quantification of an open sentence, P(x), in the following way:

$$\forall x \in \mathrm{U}, \mathrm{P}(x).$$

For instance, let P(x) be "$x + 2 = 2 + x$." Then, $\forall\ x \in \mathrm{W}$, P($x$) symbolizes the statement, "For every whole number x, $x + 2 = 2 + x$." For any open sentence P(x) and an associated universal set U, $\forall\ x \in \mathrm{U}$, P($x$) is not an open sentence but a statement.

Notice that in the preceding example $x + 2 = 2 + x$ is a true statement for each replacement of x by a whole number from W. Hence, $\forall\ x \in \mathrm{W}$, P($x$) is a true statement. For a universally quantified statement to have a truth value of T, it must be the case that, when the variable is replaced with each element of the universal set, the resulting statement is true.

We call the quantifier "there exists" the *existential quantifier* and denote it by the symbol $\exists$. The quantification symbol $\exists$ is used before a variable, as $\exists x$. Let U be a universal set.

The expression $\exists x \in U$ represents several different phrases that have essentially the same meaning. Some of the more common phrases that $\exists x \in U$ represents are:

"There exists an x element of U such that"

"There is at least one x element of U such that"

"For some x element of U"

We symbolize the existential quantification of an open sentence, $P(x)$, in the following way:

$$\exists x \in U, P(x).$$

For example, let $U = \{a, b, c, 1, 2, 3\}$ and let $P(x)$ be "x is a letter of the English alphabet." Then we can symbolize the statement, "there exists an x element of U such that x is a letter of the English alphabet," by $\exists x \in U, P(x)$.

To show that $\exists x \in U, P(x)$ is true, we need to exhibit only one element of U, say a, such that P(a) is true. It doesn't matter if more than one element of the universal set makes $P(x)$ true or if some of the elements of U make $P(x)$ false as long as there is one element of U that makes $P(x)$ true. If there is one element of U that makes $P(x)$ true, then $\exists x \in U, P(x)$ is a true statement.

When we specify a universal set and prefix a universal quantifier to an open sentence, we convert the sentence into a statement. Similarly, if we specify a universal set and prefix an existential quantifier to an open sentence, we convert the sentence into a statement.

Sometimes in mathematics quantifiers are omitted by the writer, and the reader is left to supply the quantifier that will convert the open sentence into a *true* statement. Consider the open sentence "$3x + 2 = 2 + 3x$." The universal quantifier, $\forall$, will make this sentence true for the universal set W. We notice that the existential quantifier, $\exists$, will also make "$3x + 2 = 2 + 3x$" true for the universal set W. In such cases, we agree that the universal quantifier is the understood quantifier.

In summary, we see that we have two methods for obtaining a statement from an open sentence:

1) Replace the variable with a specific element from a universal set.

2) Quantify the variable and specify a universal set.

The authors have found that if students will associate "universal" with the word "all" and the upside down A ($\forall$) and "existential" with the word "exists" and the backwards E ($\exists$), it will help prevent confusing the symbols.

Exercises 2.2

1. Indicate whether the following are open sentences (O), statements (S), or neither open sentences nor statements (N).
a) There is a ship which can carry 1200 people.
b) Wait a few minutes, John.
c) For all x, element of U, x is a national park in the United States. U is the set of parks in the United States.
d) There exists an x, element of U, such that x is one of the continents of the earth. U = {Africa, Japan, North America}.
e) Which one of these topics seems the most interesting?
f) There is a course in this school that is fascinating to me.
g) x must ride on stand-by on Flight 437.
h) Some of the books in the library are out of print now.
i) x is the capital of a European country, and x is a city of over 100,000 people.
j) She and I toured x together.
k) "Come along with me to the wild country."
l) For every x, element of U, and for some y, element of U, $x + y = 13$. U is the set of integers Z.

2. Let U = {1,2,3,...,9}. Let P(x) be $4 \cdot x = 36$. What is
a) P(2)
b) P(5)
c) P(6)
d) P(9)

What are the truth values for (a), (b), (c) and (d)?

3. Let U = $\{x \mid x$ is a color$\}$. Let P(x) be the sentence, "x is a color in the U.S. flag." What is
a) P(red)
b) P(yellow)
c) P(blue)
d) P(black)

What are the truth values for (a), (b), (c) and (d)?

4. Let U = {1789, 1770, 1754, 1773}. Let P(x) be the

sentence, "x is the year of the Boston Tea Party, or x is the year of the Boston Massacre." What is

a) P(1789) b) P(1770) c) P(1754) d) P(1773)

What truth values would you assign to a), b), c) and d)?

5. Let P(x) be the sentence, "x is the name of a city with a population of more than 400,000 people."

a) Let U = {Houston, Kansas City, Milwaukee}. What is,

1) P(Houston) 2) P(Milwaukee)

What is the truth value for 1)? for 2)?

b) Let U = {Tucson, Reno, Hattiesburg, Conway}. What is,

1) P(Reno) 2) P(Tucson)

What is the truth value for 1)? for 2)?

c) Name two additional sets that could serve as replacement sets for P(x).

6. Let U = {beneficial, profitable, helpful, serviceable}. Let P(x) be "x is a word containing 10 different letters of the English alphabet."

a) Is $\exists x \in$ U, P(x) true or false? Why?

b) Is $\forall x \in$ U, P(x) true or false? Why?

7. Let P(x) be the sentence, "x is a musical instrument."

a) Specify a replacement set for P(x) and then convert P(x) into a statement by replacing the variable with an element from U.

b) Specify a replacement set for P(x) and then convert P(x) into a statement by using the existential quantifier.

c) Specify a replacement set for P(x) and then convert P(x) into a statement by using the universal quantifier.

8. True (T) or False (F).

a) Any open sentence can be converted into a statement.

b) Any sentence can be converted into a statement.

c) If $\exists x \in$ U, P(x) is a statement, then P(x) must be a statement.

d) The universal quantifier is symbolized by $\forall$.

e) A set of possible replacements for a sentence variable is called a universal set for that sentence.

f) The quantifier $\exists x \in$ U is interpreted, "There exists an x, element of U."

g) Open sentences are statements that contain a variable.

h) The existential quantifier is symbolized by $\exists$.
i) If a is an element from the replacement set for the open sentence $P(x)$, then P(a) is a statement.

9. Let $P(x)$ be "x is a fruit that is grown in California and in Florida."
a) Specify three sets that can serve as replacement sets for $P(x)$.
b) Suppose $\exists x \in U, P(x)$ has a truth value of T.

Then, how would a quantified statement having the truth value F be symbolized?

Section 3

Venn Diagrams and Statements

Our preceding examples have dealt primarily with simple statements. There are some interesting questions that arise when one starts dealing with compound statements. Two natural questions are:

1. How do we compound statements?
2. How do we determine the truth value of a compound statement if we know the truth values of its components?

In the process of answering these questions, we will present some additional terminology and notation. In connection with our studies in logic, the variables we will use will be placeholders for statements. The variables are designated by small letters, p, q, r, etc. We will allow statement variables to be placeholders for simple as well as for compound statements. Do not confuse this notation for variables with the previously introduced notation for a particular statement. For example, if we are designating the statement "All dogs are animals" by a letter, we would use a capital letter, say P. When we use a small letter, like p, we are letting p be a placeholder for an unspecified statement.

There is a close connection between much of the set theory we studied in Chapter 1 and the logical relationships we will develop in this chapter. We will use our familiarity with sets to help illuminate the path in this development.

Let P be the statement "Today is Wednesday." Notice

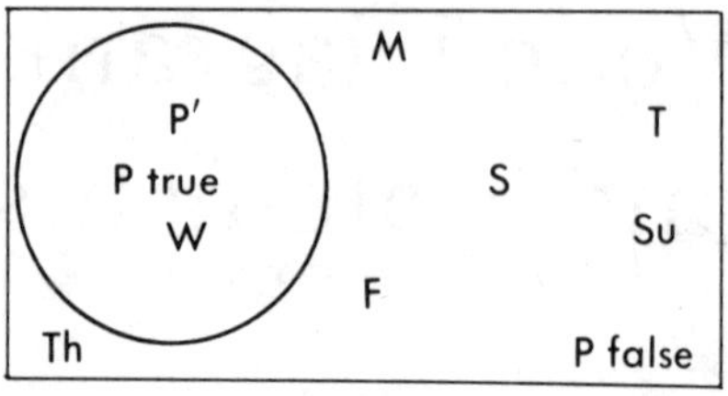

Figure 2.1

that the set of all possible days of the week is {Monday (M), Tuesday (T), Wednesday (W), Thursday (Th), Friday (F), Saturday (S), Sunday (Su)}. Consider the Venn diagram in Figure 2.1. Let U be the days of the week, and let P′ be the subset of U consisting of those days for which statement P is true. Thus, P′ = {Wednesday}.* Figure 2.1 helps us visualize the set U of all logical possibilities for statement P (all conditions under which P is either true or false). It also shows P′, the subset of U consisting of all the conditions for which P is true.

We can generalize the above discussion in the following way. Let p represent a statement and let U represent the set of all logical possibilities for p (all conditions under which p is either true or false). Let P′ be the subset of U representing the logical possibilities for which p is true. P′ is called the truth set of p. The Venn diagram in Figure 2.2 represents this condition. Notice that the shaded area, P′, represents "those conditions for which p is true." That is, P′ represents the truth set of p. Those objects in U but outside of P′ represent conditions under which p is false.

Suppose there are no logical possibilities for which p is false. Then, P′ = U, and we say that p is *logically true*. The

*We shall consider statement P to be uttered on Wednesday. It is clear that changing the context of the statement would possibly change its truth value; that is, changing the day on which the statement was uttered would possibly change its truth value. In general, the reader is cautioned that changing the context of a statement may render the statement ambiguous or may change its truth value.

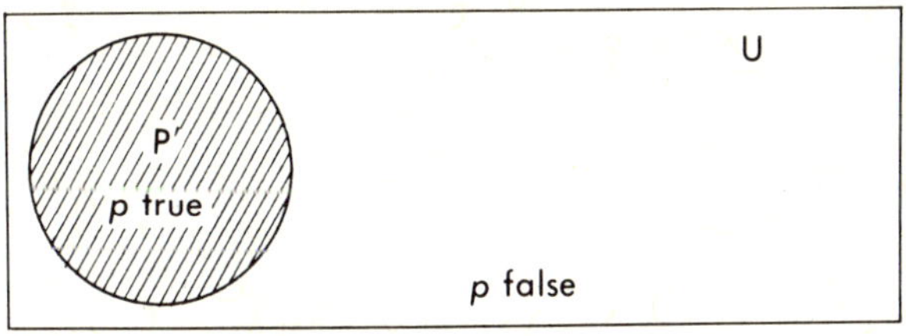

Shaded area represents truth set for p.

Figure 2.2

statement "Today is one of the seven days of the week" is a logically true statement. That is, none of the logical possibilities for the statement makes the statement false. Similarly, if p is not true for any logical possibilities, then $P' = \phi$, and we say p is *logically false.* If p is not logically true or logically false, then we say p is *contingent.* Notice that the p in the diagram in Figure 2.2 represents a contingent statement. Unless noted otherwise, statement variables will not be placeholders for simple statements that are logically true or logically false.

The ideas just expressed are very important. They will be the basis for our development of symbolic logic through set-theoretic concepts. We have introduced several symbols to aid us in our work. The chart in Figure 2.3 will be helpful in distinguishing among these symbols.

Symbols	Explanation
P, Q, R, etc.	Represent specific statements
p, q, r, etc.	Placeholders for unspecified statements
$P(x), Q(x), R(x)$, etc.	Represent open sentences in variable x
P(a), Q(a), R(a), etc.	Represent statements formed from open sentences whose variable is replaced by a
P', Q', R', etc.	Represent truth sets for statements p, q, r, etc.

Figure 2.3

Section 4
Negation

We frequently negate statements in our daily conversation. In negating a statement we can prefix it with the phrase "It is false that" or the phrase "It is not the case that." For example, let P be the statement "Every dog has his day." The negation of P, denoted by $\bar{P}$, can be expressed in any of the following ways.

a) Not every dog has his day.
b) It is not the case that every dog has his day.
c) There exists a dog that does not have his day.
d) Some dog does not have his day.

The negation of any true statement is false, and the negation of any false statement is true. The negation of a statement variable p is denoted by $\bar{p}$.

Let U be the set of logical possibilities for p and P′ the subset of U for which p is true. Now $\overline{P}'$ represents all logical possibilities for which p is false. $\overline{P}'$ also represents all the logical possibilities for which $\bar{p}$ is true, and P′ represents the subset of U for which $\bar{p}$ is false. The above relationships may be visually represented by a Venn diagram, as in Figure 2.4. The table given in Figure 2.4 summarizes the possibilities illustrated in the Venn diagram. We call such a table a *truth table*. The numerals are used to designate those regions in the Venn diagram that correspond with the truth values in the truth table.

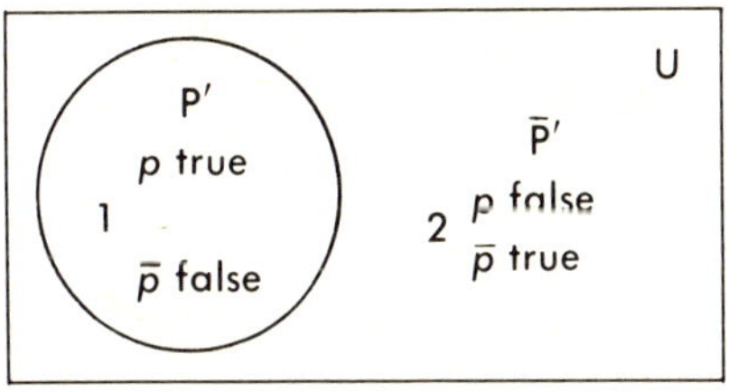

	p	$\bar{p}$
1	T	F
2	F	T

Figure 2.4

Many times it is a straightforward process to determine the wording for a negated statement. However, for a novice it can be confusing at times to know how to express the negation of a statement. (It can also be confusing to the experienced if the statement is complex enough.)

Consider the statement P to be, "It happened at dawn." Suppose we say that $\bar{P}$, the negation of P, is, "It happened at dusk." Now, if it happened at noon, both P and $\bar{P}$ would be false. This violates our rule for negating statements, which says that if P is false then $\bar{P}$ must be true. Thus, $\bar{P}$ cannot be, "It happened at dusk." A correct negation of statement P is, "It did not happen at dawn."

Let $P(x)$ denote the open sentence "x is an even number," and let the universal set be $\{2,4,6,8\}$. For each x element of U, we see that $P(x)$ is true. Hence, we can say that the statement $\forall\, x \in U,\ P(x)$ is true. It would be incorrect to say that there exists a number in the universal set that is not even. We can, therefore, state that $\exists x \in U,\ \overline{P(x)}$ is false. We can also say that it is not the case that there exists a number in the universal set that is not even; that is, $\overline{\exists x \in U,\ \overline{P(x)}}$, is true. Notice that $\forall\, x \in U,\ P(x)$ and $\overline{\exists x \in U,\ \overline{P(x)}}$ make the same assertion. The first statement states that for every $x \in U$, $P(x)$ is true, and the second statement says there is no $x \in U$ for which $P(x)$ is false.

We note that $\exists\, x \in U,\ P(x)$ is also true (for the same U and $P(x)$ as above), since there is at least one element of U for which $P(x)$ is true. We can also say that $P(x)$ is not false for any element in the universal set. That is, $\overline{\forall x \in U,\ \overline{P(x)}}$ is also a true statement. Notice that $\exists x \in U,\ P(x)$ means,

"There is one $x \in \mathrm{U}$ such that $\mathrm{P}(x)$ is true," and $\overline{\forall x \in \mathrm{U}, \overline{\mathrm{P}(x)}}$ means, "It is not the case that for each $x \in \mathrm{U}$, $\mathrm{P}(x)$ is a false statement."

So that the happiness quotient of the student may remain high, the authors state the following guides for negating quantified statements.

1. The negation of $\forall x \in \mathrm{U}, \mathrm{P}(x)$ is $\exists x \in \mathrm{U}, \overline{\mathrm{P}(x)}$.
2. The negation of $\exists x \in \mathrm{U}, \mathrm{P}(x)$ is $\forall x \in \mathrm{U}, \overline{\mathrm{P}(x)}$.

Exercises 2.3, 4

1. Give an example of a logically true statement and an example of a logically false statement.

2. Negate the following statements.

a) This is not a very elaborate building.
b) I have nothing to do during my spare time.
c) The fence is white.
d) The business transaction was unprofitable.
e) Because the thief was running so fast, the police were not able to distinguish his features.
f) Children are happy people.
g) We are going to the drugstore, and then we are going to the ice rink.
h) The tablecloth was circular and had a floral design in its center.
i) My parents were unable to convince me that I had made a mistake.
j) Oliver is a dog, or else he is a cat.

3. Let $\mathrm{P}(x)$ be "x is a money-making enterprise." Let U be $\{x \mid x \text{ is a record club}\}$. Write the following in words.

a) $\forall x \in \mathrm{U}, \mathrm{P}(x)$
b) $\exists x \in \mathrm{U}, \mathrm{P}(x)$
c) $\exists x \in \mathrm{U}, \overline{\mathrm{P}(x)}$
d) $\forall x \in \mathrm{U}, \overline{\mathrm{P}(x)}$
e) $\overline{\forall x \in \mathrm{U}, \mathrm{P}(x)}$
f) $\overline{\exists x \in \mathrm{U}, \mathrm{P}(x)}$
g) $\overline{\exists x \in \mathrm{U}, \overline{\mathrm{P}(x)}}$
h) $\overline{\forall x \in \mathrm{U}, \overline{\mathrm{P}(x)}}$

4. Let $\mathrm{P}(x)$ be "x is an interesting place," and let U be $\{x \mid x \text{ is a bookstore}\}$. Write the following in words.

a) $\forall x \in \mathrm{U}, \mathrm{P}(x)$
b) $\exists x \in \mathrm{U}, \mathrm{P}(x)$
c) $\exists x \in \mathrm{U}, \overline{\mathrm{P}(x)}$
d) $\forall x \in \mathrm{U}, \overline{\mathrm{P}(x)}$
e) $\overline{\forall x \in \mathrm{U}, \mathrm{P}(x)}$
f) $\overline{\exists x \in \mathrm{U}, \mathrm{P}(x)}$
g) $\overline{\exists x \in \mathrm{U}, \overline{\mathrm{P}(x)}}$
h) $\overline{\forall x \in \mathrm{U}, \overline{\mathrm{P}(x)}}$

5. Write each of the following quantified statements in symbolic form. Then, write in words the negation of the statement.

a) Every dress I own needs to have its hem raised.

b) There is at least one term paper that deserves an A.

c) Some of the watermelons were not ripe.

d) Not everyone can be outstanding in sports.

e) There are some good novels being written by current American authors.

f) There exists a river in Arizona that has water in it.

g) All of the applicants were not well qualified.

h) Every question I asked was ignored.

i) All cats are furry and sleek.

j) Each local garden club has its own flower show.

Section 5
Conjunction

The word *and* is used to connect simple statements. For instance, let P symbolize "Long hair is attractive," and let Q symbolize "Long hair requires a lot of care." We can compound these statements into the statement "Long hair is attractive, and long hair requires a lot of care." By using symbols, we can shorten this statement to "P and Q." The statement P and the statement Q are connected by the word *and*. The resulting statement is called the *conjunction* of statements P and Q. Thus, conjunction might be considered a binary operation on statements. In daily discourse, the word *and* is used for more than just connecting statements as we have defined them. For instance, the expression "Bill and Joe are brothers" uses *and* to express a relationship and not to connect two statements. We will use the symbol "$\wedge$" to denote the operation of conjunction of statements. Thus, we symbolize the conjunction, "Long hair is attractive, and long hair requires a lot of care," by $P \wedge Q$.

Let p and q be statement variables. We must decide on the truth value for $p \wedge q$ by examining the truth value possibilities of p and q separately. For statement variables p and q, we have the possibility that p is either true or false and we have the possibility that q is either true or false. There are four possible combinations of truth values for the statement variables in $p \wedge q$. They are as follows:

a) $p \wedge q$ b) $p \wedge q$ c) $p \wedge q$ d) $p \wedge q$
    T T   T F   F T   F F

How can we decide on the appropriate truth values for $p \wedge q$ in each of these cases? Let us consider our previous statement, "Long hair is attractive, and long hair requires a lot of care," symbolized by P ∧ Q. We agree P ∧ Q is true only if it is true that "Long hair is attractive" and if it is true that "Long hair requires a lot of care"; that is, only if P is true and if Q is true. If at least one of the component statements is false, then P ∧ Q is false. This example shows the plausibility of assigning a truth value of T to $p \wedge q$ only when p is true and when q is true. In the other three cases, $p \wedge q$ is false.

The operation of intersecting two sets is related to the connective ∧. Let p and q be statement variables. Let U represent the set of logical possibilities for p and q. Let P′ and Q′ represent the truth sets for p and q, respectively; that is, represent the subsets of U that contain all of the logical possibilities for which p and q are true. The intersection of P′ and Q′ represents the set consisting of the logical possibilities for which p is true and q is true. Hence, P′ ∩ Q′ represents the truth set of $p \wedge q$. This is illustrated in the Venn diagram in Figure 2.5. The truth table for $p \wedge q$ is given in Figure 2.6. Numerals are used to designate those

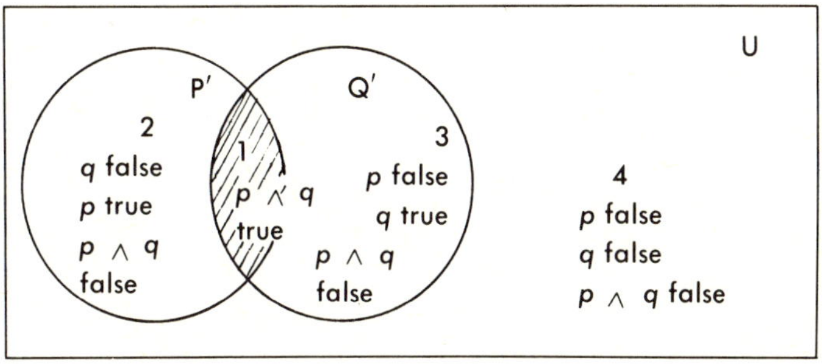

Truth set for $p \wedge q$ shaded in

Figure 2.5

	p	q	$p \wedge q$
1	T	T	T
2	T	F	F
3	F	T	F
4	F	F	F

Figure 2.6

regions in the Venn diagram that correspond with the truth values in the table.

There are several words in English that can express the conjunctive sense of *and*. Examples of the words are *yet, but, moreover, however,* and *although*. When these words are used in their conjunctive sense, we can symbolize them by $\wedge$.

Section 6
Disjunction

For full-time employment, many employers require that "An applicant must be a graduate of the eighth grade, or an applicant must be over sixteen years of age." This requirement illustrates the use of another connective, "or," which connects the simple statements "An applicant must be a graduate of the eighth grade" and "An applicant must be over sixteen years of age." Let P stand for "An applicant must be a graduate of the eighth grade," and let Q represent "An applicant must be over sixteen years of age." Using these symbols, we can express an employer's requirement as "P or Q." The statement "P or Q" is called the *disjunction* of statements P and Q. The symbol " $\vee$," read *or,* is used to denote the binary operation of disjunction on statements. Unless otherwise stated, the word *or* will be used in its inclusive sense, as in the above case. Recall that this is the same manner in which *or* was used in Chapter 1.

Let p and q be statement variables. We must decide on the truth value for $p \vee q$ by examining the truth value possibilities of p and q separately. As before, we have four possible cases to consider. There are two possible truth values for p, T and F, and independent of each of these, q has two possible truth values, T and F. Since we are using the inclusive *or,* we must agree that $p \vee q$ is true if p is true and if q is true. In our example above, if an applicant is a graduate of the

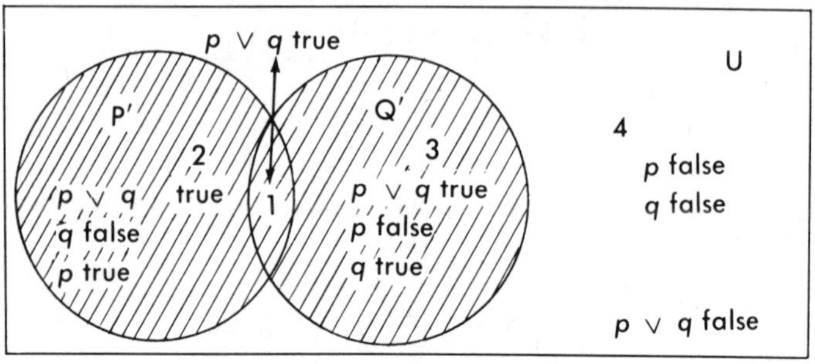

Truth set for $p \vee q$ shaded

Figure 2.7

eighth grade and is over sixteen years of age, it is undoubtedly true that he satisfies the requirement for full-time employment. In agreement with everyday usage of the inclusive sense of or, $p \vee q$ is true if either p is true or q is true. If both p and q are false, we certainly feel comfortable in assigning $p \vee q$ a truth value of false.

As the set operation of intersection is related to the connective *and,* so the set operation of union is related to the connective *or.* Let U represent the set of logical possibilities for statement variables p and q. Let P′ and Q′ represent the subsets of U that contain all of the logical possibilities for which p and q are true, respectively. The union of sets P′ and Q′, P′ ∪ Q′, represents the truth set for $p \vee q$. This is illustrated in the Venn diagram in Figure 2.7. The truth table for the disjunction $p \vee q$ is given in Figure 2.8. As be-

	p	q	$p \vee q$
1	T	T	T
2	T	F	T
3	F	T	T
4	F	F	F

Figure 2.8

fore, we will use numerals to associate the Venn diagram for the union of sets P′ and Q′ with the truth table of $p \vee q$.

The word *unless,* used in the disjunctive sense, is also symbolized by $\vee$.

Exercises 2.5, 6

1. In each of the following, decide whether or not *and* is used in the conjunction of two statements.
a) Charlie and George work well together.
b) Both salt and pepper should always be placed on the table.
c) During my vacation, I am going to Yellowstone, and I am going to Yosemite.
d) The dictaphone is not working, and the secretary is upset.
e) Lynn and Margaret are good companions.
f) A very tasty dish results when one mixes turnip greens and mustard greens.
g) The ballpoint pen skips and occasionally will not even work.

2. In each of the following decide if *or* is used in its inclusive sense (I) or its exclusive sense (E).
a) Clowning around or talking loudly is not acceptable behavior.
b) The children tried to aggravate me or to please me according to their mood.
c) To get the job you must read German or French.
d) Soup or sandwiches would be good to eat.
e) Suzanne can be found either at school or at home.

3. Let P be "The moon is beautiful" and let Q be "It is dark outside tonight." Using the symbols P, Q, $\wedge$, $\vee$, and $^{-}$, express the following sentences symbolically.
a) The moon is beautiful, and it is dark outside tonight.
b) It is dark outside tonight, but the moon is not beautiful.
c) The moon is beautiful unless it is dark outside tonight.
d) Although it is dark outside tonight, the moon is beautiful.
e) It is not dark outside tonight; however, the moon is beautiful.

f) It is false that the moon is beautiful or it is dark outside tonight.

g) It is not the case that the moon is not beautiful yet it is not dark outside.

h) The moon is not beautiful, nor is it dark outside tonight.

4. Let P be "People are funny," let Q be "Queens are tyrants," and let R be "Rainbows are colorful." Write the following symbolic statements in words.

a) $P \wedge Q$
b) $R \vee P$
c) $\overline{Q} \wedge R$
d) $\overline{Q \vee R}$
e) $\overline{P} \vee \overline{Q}$
f) $P \wedge \overline{R}$
g) $Q \vee \overline{R}$
h) $\overline{P} \vee Q$
i) $\overline{\overline{Q} \vee \overline{P}}$
j) $\overline{P} \wedge \overline{R}$
k) $\overline{P \wedge Q}$
l) $\overline{\overline{Q} \wedge \overline{R}}$
m) $(Q \wedge P) \wedge R$
n) $Q \wedge (P \wedge R)$
o) $R \vee (Q \wedge P)$
p) $(R \vee Q) \wedge P$

5. Using the tables in Figures 2.6 and 2.8, answer the following:

a) Suppose p has truth value F and q has truth value F. What is the truth value of $p \vee q$?

b) Suppose $p \vee q$ has truth value F and p has truth value F. What do you know about the truth value of q?

c) If $p \wedge q$ has truth value T and q has truth value T, what do you know about the truth value of p?

d) If $p \wedge q$ has truth value F and p has truth value T, what do you know about the truth value of q?

e) If $p \vee q$ has truth value T and q has truth value F, what do you know about the truth value of p?

f) If $\overline{p}$ has truth value F and $\overline{q}$ has truth value T, then what is the truth value of $q \vee p$?

g) If $\overline{q}$ has truth value F and p has truth value F, then what is the truth value of $q \wedge \overline{p}$?

Section 7
Constructing Truth Tables

To handle more complex compound statements, we need to agree on some "ground rules." Let p, q, and r be statement variables. Consider the expression $p \wedge q \vee r$. Recall that to avoid ambiguities, we used parentheses to give directions for interpreting an expression involving more than two sets. We use the same technique for making the expression $p \wedge q \vee r$ precise since this expression can mean either $(p \wedge q) \vee r$ or $p \wedge (q \vee r)$. Thus, it is important that we use parentheses (brackets, braces) to clearly indicate what we mean.

We need to make explicit an idea that we have been using implicitly with regard to statement variables. Let P be "Water is wet," and let Q be "Snow is cold." Then P ∧ Q symbolizes, "Water is wet, and snow is cold." That is, P ∧ Q is a statement. We have said that small letters, such as p and q, are placeholders for unspecified statements. When we use these letters by themselves or with a connective, such as ∧ or ∨, we have *a statement form*. For example, p, $p \wedge q$, and $p \vee q$ are all statement forms. By replacing the variables in a statement form with specific statements, we get a statement that is an *instance* of that statement form. Consider the statement form $p \wedge q$. If we replace p with P and q with Q (P and Q as designated above), we get the statement P ∧ Q. We agree that the same statement must replace the same statement variable every time the variable occurs in the same statement form.

As you may have noticed, we have used small p's and q's in our truth tables. This means that we have given truth tables for statement forms. In symbolic logic, the form of a statement is pertinent while its content is not. Hence, we determine the truth table for a statement form by considering the truth value possibilities of its component statement variables.

Let P, Q, and R be the statements "Water is wet," "Snow is cold," and "Cows can fly," respectively. The compound statement, "Water is not wet and snow is cold, or cows can fly," is symbolized by $(\overline{P} \wedge Q) \vee R$. What is the truth value of this statement? In determining the truth value of $(\overline{P} \wedge Q) \vee R$, we will give a technique for determining a truth table for the statement form $(\overline{p} \wedge q) \vee r$ of which $(\overline{P} \wedge Q) \vee R$ is an instance.

We notice that there are three statement variables in the statement form $(\overline{p} \wedge q) \vee r$. We must consider all possible truth value combinations for these three variables. You will remember that for two variables we had to consider four possible cases. After some consideration, you will see that for three variables we need to consider eight possible cases.*

Figure 2.9 shows a convenient way to arrange a truth table for $(\overline{p} \wedge q) \vee r$. Notice that columns 1, 2, and 3 of the table in Figure 2.9 illustrate an easy means for obtaining all of the truth value combinations for the three statement variables.

After writing the eight possible truth value combinations in columns 1, 2, and 3, we then write column 4 ($\overline{p}$), using the rule for negation given in Figure 2.4. To determine the

*Take three pennies. Place them in the following order:

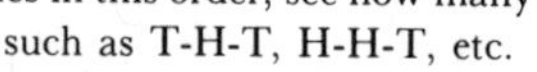

1st penny 2nd penny 3rd penny

Now, leaving the pennies in this order, see how many combinations of heads and tails you can get, such as T-H-T, H-H-T, etc. If you let heads stand for T and tails stand for F, you will have the same number of combinations of T and F as you have of heads and tails.

appropriate truth values for column 5 ($\bar{p} \wedge q$), we use first column 4 and then column 2, along with the rule for conjunction given in Figure 2.6. Notice that the parentheses give us directions as to what symbols to group together. Column 6 is the result of using first column 5 and then column 3, along with the rule for disjunction given in Figure 2.8. Now, column 6 is the final column of our table. It is the last column that tells us the truth values for the statement form $(\bar{p} \wedge q) \vee r$ under different combinations of truth value possibilities of the component statement variables.

What is the truth value of the statement $(\bar{P} \wedge Q) \vee R$? First, we must determine the truth values of the component statements. P is "Water is wet," which is true; Q is "Snow is cold," which is true (at least this is the authors' experience); and R is "Cows can fly," which is false. Hence, the truth values for P, Q, and R are T, T, and F, respectively. Notice that row 2 on our truth table has this same combination of truth values (T, T, F) listed in columns 1, 2, and 3, respectively. Moving along this row to the last column, column 6, we find an F. Hence, the statement $(\bar{P} \wedge Q) \vee R$ is false, F. The other seven rows indicate the truth or falsity of the statement form under the other possible truth value combinations specified in columns 1, 2, and 3 of the corresponding row.

The table in Figure 2.9 shows all of the possible combi-

Columns

	1	2	3	4	5	6
	p	q	r	$\bar{p}$	$\bar{p} \wedge q$	$(\bar{p} \wedge q) \vee r$
1	T	T	T	F	F	T
2	T	T	F	F	F	F
3	T	F	T	F	F	T
4	T	F	F	F	F	F
5	F	T	T	T	T	T
6	F	T	F	T	T	T
7	F	F	T	T	F	T
8	F	F	F	T	F	F

Figure 2.9

Columns

	1	2	3	4
	p	q	$\bar{p}$	$\bar{p} \vee q$
1	T	T	F	T
2	T	F	F	F
3	F	T	T	T
4	F	F	T	T

Figure 2.10

nations of truth values for the component statement variables of the statement form $(\bar{p} \wedge q) \vee r$. Thus, if we desire to know the truth value of a particular statement that is an instance of the statement form $(\bar{p} \wedge q) \vee r$, all we do is find the row in which the entries for p, q, and r match the truth values of the component statements P, Q, and R. For example, suppose our truth values for P, Q, and R were F, T, and F, respectively. Then we would look at row 6. Going across row 6 to column 6, we see that in this case, $(\bar{P} \wedge Q) \vee R$ has the truth value of T.

To see another truth table, look at Figure 2.10. The table in this figure gives the truth values for the statement form $\bar{p} \vee q$ under the various combinations of truth values of the component statement variables p and q.

To build the table in Figure 2.10, put all of the possible truth value combinations in columns 1 and 2. Column 3 follows by negating column 1. Finally, use the rule for disjunction along with columns 3 and 2, in that order, to get column 4.

There are some general comments that need to be made regarding the procedure for developing truth tables.

Figure 2.11 indicates a general pattern for determining the number of truth value combinations when the number of statement variables is known. Notice that $2^1 = 2$, $2^2 = 4$, $2^3 = 8$, where the number that is the superscript (exponent) is the number of statement variables. Refer to the notation introduced in Exercises 1.2, 3, Problem 4. If n is the number

No. of Statement Variables	No. of Truth Value Combinations
1	2
2	4
3	8
.	.
.	.
.	.
n	2^n

Figure 2.11

of statement variables, then 2^n is the number of possible truth value combinations.

There is a mechanical procedure for making sure that all of the possible combinations of truth values for the statement variables are included. Suppose you have n statement variables. To make sure you include all of the possible truth value combinations for the n statement variables, proceed as follows:

1) List these variables in order, using p, q, r, etc., in the first n columns of the table.

2) In the nth column, the first entry should be T. The next entry should be F. Continue alternating T and F until you have written 2^n truth values.

3) Begin at the top of the next column to the left and list twice as many Ts as initial Ts in the column to the immediate right. Follow this with the same number of Fs and continue alternating this number of Ts and Fs until you have written 2^n truth values.

4) Continue to repeat the procedure outlined in step 3 above until you have completed all n columns. Notice that in each column you have the same number of Ts and Fs, but in each column there is a different arrangement of these truth values.

Truth tables are an important tool, and you need to be sure you know how to build them before proceeding further. There are two basic rules that you need to remember when building a table. First, work from the inside out when dealing with complex compound statements, which have parentheses, braces, and brackets. Second, the truth value possibilities of the compound statement form under consideration are found in the last column of the table.

It is interesting to note that we are now able to formulate an expression for the exclusive *or*. The expression is $(p \vee q) \wedge (\overline{p \wedge q})$. You should build a truth table for this statement form and compare it with the table for the inclusive *or*, Figure 2.8.

Exercises 2.7

1. Construct truth tables for the following statement forms.

a) $(p \wedge q) \vee \bar{r}$
b) $(p \vee q) \wedge (p \wedge r)$
c) $p \wedge \bar{q}$
d) $\overline{p \vee q}$
e) $\bar{q} \wedge p$
f) $(\overline{p \vee q}) \vee r$
g) $\bar{q} \wedge (\bar{p} \vee r)$
h) $(q \wedge r) \vee (p \wedge r)$
i) $(p \wedge r) \wedge q$
j) $p \vee [r \wedge (\bar{q} \vee r)]$
k) $[(q \wedge \bar{p}) \vee (r \wedge \bar{p})] \vee p$
l) $[\bar{q} \vee (p \vee r)] \vee \bar{r}$
m) $[(p \wedge r) \vee \bar{q}] \wedge [(\bar{q} \vee r) \wedge p]$
n) $(r \vee q) \wedge \{\bar{p} \vee [q \wedge (\overline{p \wedge r})]\}$

2. Give the symbolic statement form for each of the following statements.

a) Although my feet hurt and I have a headache, I will go to this evening's meeting of the bridge club.
b) The board of trustees had a reputation of being harsh and demanding, but they accepted the new president's program enthusiastically.
c) Men fear death or illness, and sometimes they become neurotic about their fears.
d) She was arrested and was taken to the courthouse, yet she had a smile on her face, and her countenance was calm.
e) Mr. Stone was neither amused nor happy unless his political advisor had made proper flight arrangements and had scheduled conferences at appropriate intervals.
f) I will not be happy and you will not be happy unless you make the right decision and act upon that decision.
g) They had both been through threats and bombings and had learned to hold their heads high or suffer additional ridicule.
h) Sammy went to the lake, and Steve went riding on his

motorcycle, but neither of them thought about their obligations at home.

3. Give the truth value for each of the following statement forms if p is true, q is false, and r is false.
a) $\overline{p} \wedge (\overline{\overline{q} \vee \overline{r}})$
b) $(\overline{p \vee q}) \wedge (r \vee q)$
c) $(p \wedge r) \vee q$
d) $[(r \vee \overline{q}) \wedge p] \vee (p \wedge q)$
e) $(p \vee \overline{p}) \wedge q$

4. Using the truth tables, answer the following questions.
a) If p is T and $\overline{p} \vee \overline{q}$ is F, then what is the truth value of q?
b) If q is T and $\overline{\overline{p} \vee \overline{q}}$ is F, then what is the truth value of p?
c) If q is F and $\overline{\overline{p} \vee \overline{q}}$ is T, then what is the truth value of p?
d) If $(p \wedge \overline{q}) \vee p$ is F, then what is the truth value of p and what is the truth value of q?
e) If $(p \wedge \overline{q}) \vee p$ is T and $\overline{p}$ is F, then what is the truth value of q?
f) If $\overline{q} \wedge (p \vee \overline{p})$ is T, then what is the truth value of q?
g) If $(\overline{p} \vee q) \wedge (\overline{q} \vee p)$ is T and q is F, then what is the truth value of p?
h) If $(p \wedge \overline{q}) \vee r$ is T, then what is the truth value of q?
i) If r is T and $(\overline{r} \vee q) \vee (p \wedge r)$ is F, then what is the truth value of p?
j) If $(\overline{p \wedge q}) \vee [r \wedge (\overline{q} \vee p)]$ is F, then what are the truth values of p, q, and r, respectively?

5. Build a truth table for the statement form

$$(p \vee q) \wedge (\overline{p \wedge q})$$

Note that this statement form expresses the exclusive *or*. Compare the table for the inclusive *or* (Figure 2.8) with the table you just constructed.

6. Write in words the following symbolic statements. Then find their truth values, given that P is "Dogs cannot fly" and has truth value T, Q is "Birds cannot swim" and

has truth value T, and R is "Fish can bark" and has truth value F.

a) $P \wedge \overline{Q}$
b) $\overline{P \wedge Q}$
c) $Q \vee \overline{R}$
d) $\overline{P} \wedge Q$
e) $(\overline{P} \wedge \overline{Q}) \vee R$
f) $(P \vee R) \wedge (\overline{R} \vee P)$
g) $P \vee Q$
h) $\overline{P} \vee \overline{R}$
i) $\overline{\overline{P} \wedge \overline{Q}}$
j) $\overline{\overline{Q} \vee \overline{R}}$

7. How many cases does one need to consider when constructing the truth table for a statement form with

a) 3 variables?
b) 4 variables?
c) 5 variables?
d) 6 variables?

Section 8

Tautologies and Contradictions

The truth table for the statement form $p \vee \bar{p}$ is shown in Figure 2.12. Notice that no Fs occur in the final column of the truth table. This means that $p \vee \bar{p}$ is true, regardless of the truth values of its components.

A compound statement form is said to be *logically true* if it is always true, regardless of the truth assignments of the statement variables that compose it. Hence, $p \vee \bar{p}$ is logically true. A logically true statement form is also called a *tautology*. A tautology will have only Ts in the last column of the truth table for the statement form in question. A statement that is an instance of a tautologous statement form is called a *tautologous statement*. Any statement form that has both Ts and Fs in the final column of its truth table that are dependent upon the truth assignments of the statement variables in the form is called a *contingent statement form*. A statement that is an instance of a contingent statement form is called a *contingent statement*. As you might expect, there are statement forms that have all Fs in the truth table under the statement form being considered. These statement forms are said to be *logically false* or are sometimes called *contradictions*. A statement that is an instance of a contradiction is called a *contradictory statement*. An example of a contradiction is the statement form $p \wedge \bar{p}$, as indicated in the truth table in Figure 2.13.

p	$\bar{p}$	$p \vee \bar{p}$
T	F	T
F	T	T

Figure 2.12

p	$\bar{p}$	$p \wedge \bar{p}$
T	F	F
F	T	F

Figure 2.13

As mentioned earlier, the universal set is the truth set that is associated with a logical truth. Consider the statement form $p \vee \bar{p}$. Let P′ be the subset of U for which p is true, and let $\overline{\text{P}}'$ be the subset of U for which p is false. The Venn diagram in Figure 2.14 illustrates that $p \vee \bar{p}$ has U as its truth set.

We have noted that the empty set is the truth set that is associated with a logical falsehood. Let P′ be the subset of U for which p is true. Consider the statement form $p \wedge \bar{p}$. We see from the Venn diagram in Figure 2.15 that ϕ is the truth set for $p \wedge \bar{p}$.

Tautologies and contradictions play an important role in mathematics. We will be using and extending these concepts in the rest of this chapter and in following chapters.

Exercises 2.8

1. What is the truth set for a tautology?
2. What is the truth set for a contradiction?
3. True or False

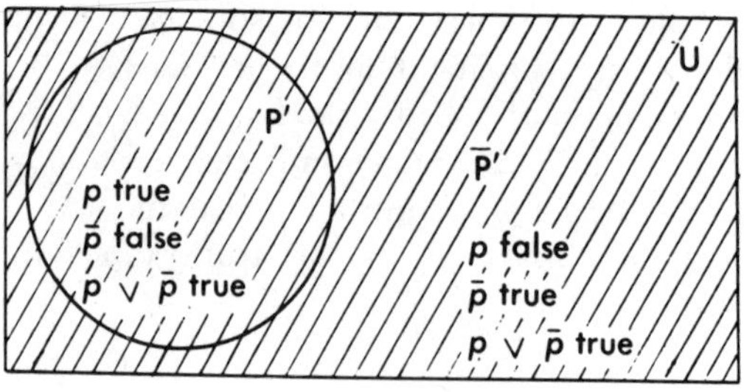

Truth set for $p \vee \bar{p}$ is shaded.

Figure 2.14

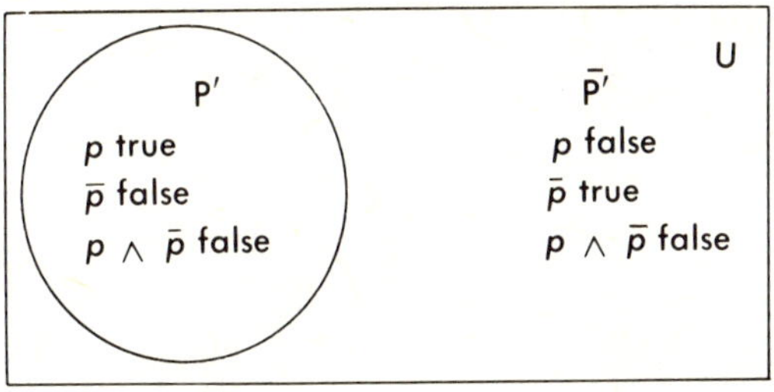

There is not a point in U for which $p \wedge \bar{p}$ is true.

Figure 2.15

a) A logically true statement is true only when each of its component variables is true.
b) A logically false statement is false only when each of its component variables is false.
c) A contingent statement is a statement that is neither a tautology nor a contradiction.
d) A logically true statement is true only when one of its component variables is true.
e) A statement that is not logically true is a logically false statement, and vice versa.
f) A logically false statement is false only when one of its component variables is false.

4. Construct the truth tables for the following statement forms, and indicate whether they are tautologies (T), contradictions (C), or contingent statement forms (CSF).

a) p
b) $p \vee \bar{p}$
c) $p \wedge (\bar{q} \wedge p)$
d) $(p \wedge q) \vee (\bar{p} \wedge \bar{q})$
e) $(\bar{p} \wedge \bar{q}) \vee (p \vee q)$
f) $[q \wedge (p \vee \bar{q})] \wedge \bar{q}$
g) $[(\bar{q} \wedge p) \wedge (q \vee \bar{p})] \vee (p \wedge q)$
h) $(\bar{q} \wedge p) \wedge (q \vee \bar{p})$
i) $[p \vee (\bar{p} \vee q)] \wedge [(\overline{p \wedge q}) \vee q]$
j) $[\overline{\bar{p} \wedge (q \vee \bar{q})}] \wedge (\bar{q} \wedge \bar{p})$
k) $[p \vee (q \wedge \bar{q})] \vee (\bar{q} \wedge \bar{p})$

Section 9
Conditional Statements

During baseball season on most college campuses, many students become quite involved in the win-loss records of their teams. In his enthusiasm, a student might make the statement, "If Clearview U loses the baseball game to Highbrow U, then I will buy you a steak dinner." To facilitate our discussion, let P be "Clearview U loses the baseball game to Highbrow U" and let Q be "I will buy you a steak dinner." Then we can express the student's statement as: "If P, then Q." A statement "If P, then Q" is an assertion with a condition. Hence, it is called a conditional statement and is symbolized by $P \rightarrow Q$. Notice that the connective, $\rightarrow$, is a binary operation on statements P and Q since $P \rightarrow Q$ is again a statement. In a conditional statement, the statement to the left of the arrow is called the *hypothesis*, and the statement to the right of the arrow is called the *conclusion*. We extend the use of these terms to conditional statement forms in the following manner: the statement form to the left of the arrow is called the *hypothesis*, and the statement form to the right of the arrow is called the *conclusion*.

Deciding on the truth value for the statement form $p \rightarrow q$ for each of the four possible cases is a little more involved than it was for the conjunction and disjunction. To help make our truth assignments for the conditional plausible, we use the statement $P \rightarrow Q$ (as indicated) for an illustration. Now, let student A be the one who makes the statement $P \rightarrow Q$, and let student B be the one who may gain a steak dinner. Under what conditions would student B feel as though student A had lied? That is, under what conditions would $P \rightarrow Q$ be false? We would no doubt all agree that, if P was true (Clearview U loses to Highbrow U) and Q was false (student A did not buy student B the steak dinner), then $P \rightarrow Q$ would be false (student B would feel that student A did not keep his word). In other words, we could say that student A tells a falsehood ($P \rightarrow Q$ is false) when "Clearview U loses the baseball game to Highbrow U and no steak dinner is bought for student B." This rephrasing can be symbolized by $P \wedge \overline{Q}$. So, if $P \wedge \overline{Q}$ is true, then $P \rightarrow Q$ is false. For $P \rightarrow Q$ to be true, we would require that $P \wedge \overline{Q}$ not be the case; that is, $P \rightarrow Q$ is true when $\overline{P \wedge \overline{Q}}$ is true. We have now reduced our consideration to some symbols with which we are already familiar.

To make our consideration general, we let p and q be statement variables and P' and Q', subsets of U, be the truth sets for p and q. Notice that $\overline{P' \cap \overline{Q}'}$ represents the truth set for the statement form $\overline{p \wedge \overline{q}}$. Figure 2.16 indicates the Venn diagram for $\overline{P' \cap \overline{Q}'}$. (You drew this diagram in problem 2(g), Exercises 1.5.) Figure 2.17 gives the associated truth table for the statement form $\overline{p \wedge \overline{q}}$.

Before we actually define the truth table for $p \rightarrow q$, we see if the truth values for the statement form $\overline{p \wedge \overline{q}}$ (see Fig-

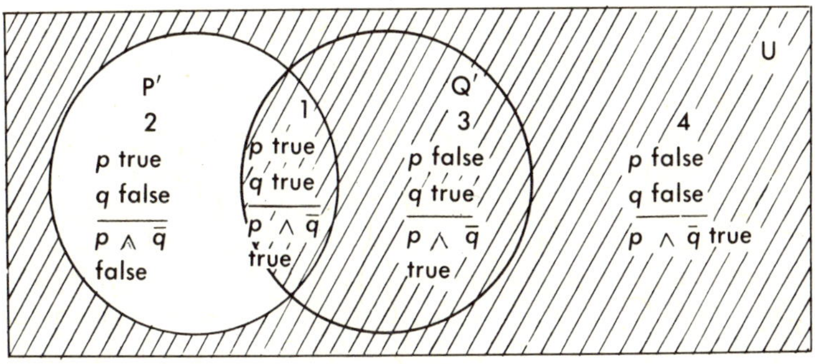

Figure 2.16

	p	q	$\overline{p \wedge \bar{q}}$
1	T	T	T
2	T	F	F
3	F	T	T
4	F	F	T

Figure 2.17

ure 2.17) are reasonable when applied to the statement form $p \rightarrow q$. Again, let P symbolize "Clearview U loses the baseball game to Highbrow U" and let Q symbolize "I will buy you a steak dinner." Let us check all four cases.

Case I: Suppose P and Q are both true; that is, "Clearview U loses the baseball game to Highbrow U, and student A buys student B a steak dinner." Student B would no doubt feel that student A told the truth; that is, P → Q is a true statement. These truth values agree with those in row 1 of Figure 2.17.

Case II: Suppose P is true and Q is false. This means, "Clearview U loses the baseball game to Highbrow U, but student A does not buy student B a steak dinner." Under these circumstances, student B would consider student A's assertion false; that is, P → Q is false. These truth values agree with those in row 2 of Figure 2.17.

Case III: Suppose P is false and Q is true. Then, "Clearview U beats Highbrow U, and student A buys student B a steak dinner." Perhaps this seems a little strange, but we would agree that student B probably would not complain or call student A a liar. Hence, we would say that P → Q is true under these circumstances, particularly since student A did not say what he would do if Clearview lost. These truth values agree with those in row 3 of Figure 2.17.

Case IV: Suppose statement P is false and statement Q is false. That is, suppose, "Clearview U beats Highbrow U, and student A does not buy student B a steak dinner." While student B might be hungry, he certainly could not call student A a liar. Thus, student A has not told a lie and P → Q is true. These truth values agree with those in row 4 of Figure 2.17.

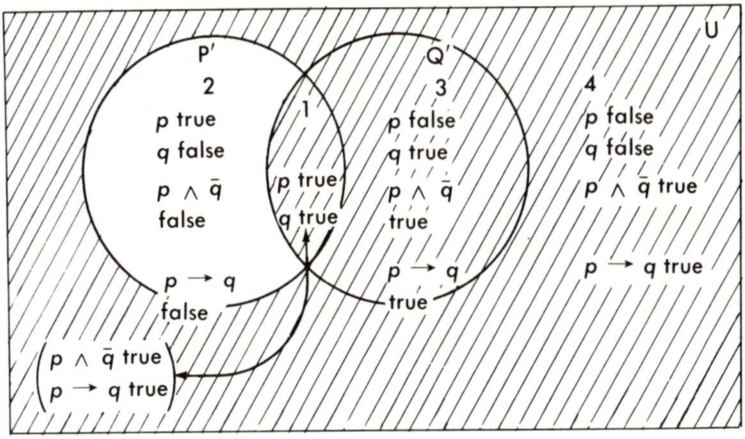

Figure 2.18

We note that in all four cases the truth values for $\overline{P \wedge \overline{Q}}$ agree with what we would intuitively expect for the conditional statement $P \rightarrow Q$. Let p and q be statement variables with associated truth sets P' and Q', subsets of U. We associate the Venn diagram in Figure 2.18 with the truth tables in Figure 2.19 and 2.20. The table in Figure 2.20 defines the truth values for the statement form $p \rightarrow q$.

In daily discourse, it is customary to use the conditional to combine statements only when they are somehow related. We make no such restriction with the conditional statement in logic. The statement, "If it is snowing in Moscow, then horseless carriages exist," is perfectly acceptable in logic. This again illustrates the fact that symbolic logic deals with the form of statements and not their content.

p	q	$\overline{p \wedge \bar{q}}$
T	T	T
T	F	F
F	T	T
F	F	T

Figure 2.19

....1....
....2....
....3....
....4....

p	q	$p \rightarrow q$
T	T	T
T	F	F
F	T	T
F	F	T

Figure 2.20

Conditional statements can be stated in many ways. The following list of statements can be phrased in the form, "If . . . , then. . . ."

1. To get to my math class on time, I must leave my home fifteen minutes before class time.
2. To succeed, one must take risks.
3. Monkey see, monkey do.
4. I will play golf this afternoon if it is not raining.
5. Infection in the body causes the white corpuscle count to increase.
6. Those that score above 90% on the final exam will receive an A.
7. You can legally drive a car, provided you are at least sixteen years of age and pass a driver's test.
8. A child will have a long and prosperous life only if he obeys his parents.

We hope you will be able to recognize a conditional statement more readily after translating these eight statements into the form "If . . . , then. . . ." To assist you further in recognizing conditional statements we list several expressions which have the form $p \rightarrow q$.

Suppose p, then q	p only if q
When p, then q	p will cause q
q whenever p	q if p
if p, then q	q provided p
p is sufficient for q	q is necessary for p

In our previous discussion, we said that a logical truth is a statement form that is true, regardless of the truth values of the component statement variables. Let p and q be statement variables. The conditional statement form $p \rightarrow q$ is called an *implication* when p and q are related in such a way that $p \rightarrow q$ is a logical truth; that is, p implies q if and only if $p \rightarrow q$ is a logical truth. Under these conditions, the truth set for the statement form $p \rightarrow q$ is U. Consider the Venn diagram associated with $p \rightarrow q$ (Figure 2.21). You will recall that we constructed this diagram in Figure 2.18. The unshaded area represents the portion of the Venn diagram for which $p \rightarrow q$ is false.

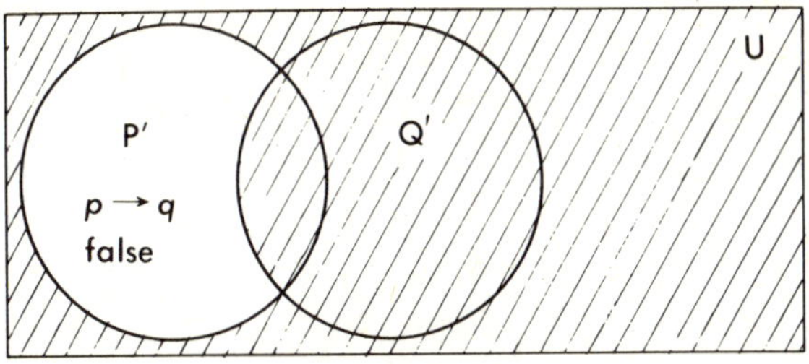

Figure 2.21

The shaded area represents the area for which $p \to q$ is true. Now, if $p \to q$ is a logical truth, then $p \to q$ is never false, and we can have no area in our universal set U representing $p \to q$ being false. That is, set P′ *must* be a subset of Q′. Figure 2.22 represents the set diagram for p implies q. Hence, we see that the relation of implication for statements is analogous to the relation of subset for sets. We use the symbol "$\Rightarrow$" to indicate an implication; that is, if $p \to q$ is a logical truth, then we can write $p \Rightarrow q$ meaning p implies q. We note that $p \Rightarrow q$ if and only if $P' \subset Q'$.

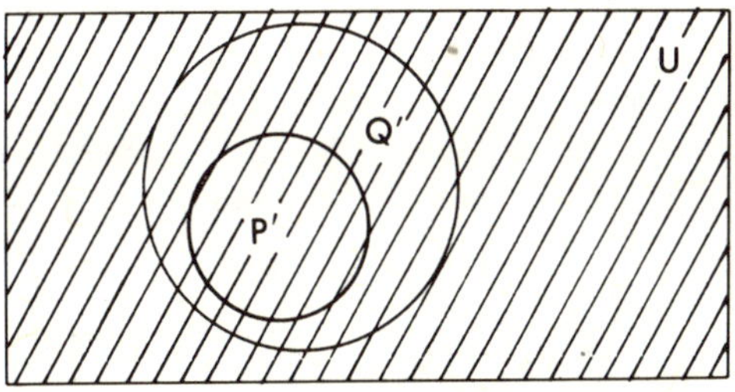

Venn diagram that shows implication, that is, p implies q.
$P' \subset Q'$
$p \Rightarrow q$

Figure 2.22

Example 4: Let p and q be statement variables. Consider the statement form $[(p \to q) \wedge p] \to q$. The truth table for this statement form is given in Figure 2.23.

Since $[(p \to q) \wedge p] \to q$ is a logical truth (tautology), we can write $[(p \to q) \wedge p] \Rightarrow q$. To show the subset relation associated with the logical truth $[(p \to q) \wedge p] \Rightarrow q$, we use Venn diagrams. Let P′ and Q′ represent truth sets for state-

p	q	$p \to q$	$(p \to q) \wedge p$	$[(p \to q) \wedge p] \to q$
T	T	T	T	T
T	F	F	F	T
F	T	T	F	T
F	F	T	F	T

Figure 2.23

ment variables p and q, respectively, and recall that $p \to q$ has truth set $\overline{P' \cap \overline{Q}'}$. Note that $[(\overline{P' \cap \overline{Q}'}) \cap P'] \subset Q'$ and, hence, $[(p \to q) \wedge p] \Rightarrow q$.

As can be seen from the previous discussion, the relation of implication has a close association to the conditional statement. To keep the concepts of "conditional statement" and "implication relation" from being confused, remember that a conditional statement is compounded from two given statements, whereas the implication relation means that a conditional statement is a logical truth.

This section on the conditional statement might seem rather involved and lengthy. We hope you have not lost sight of the concepts presented nor the inherent beauty of the ideas formulated. It might be beneficial to reread this section to clear up a few of your dangling ideas. The importance of the conditional statement cannot be overstated. *Logic* is essentially the study of "If..., then..." statements because it is with the "If..., then..." statements that we reason.

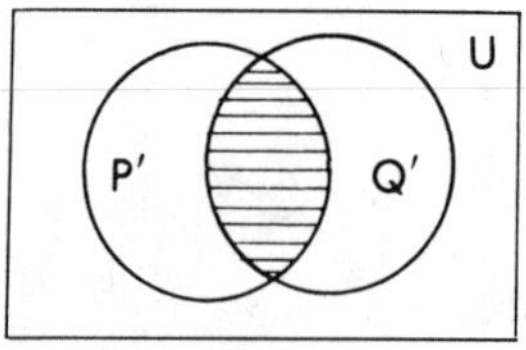

shaded area is the truth set $(\overline{P' \cap \overline{Q}'}) \cap P'$.

Figure 2.24

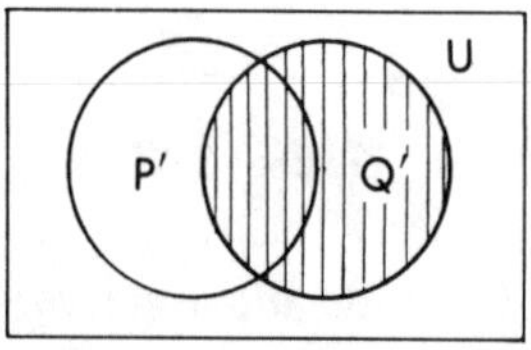

shaded area is the truth set Q'.

Figure 2.25

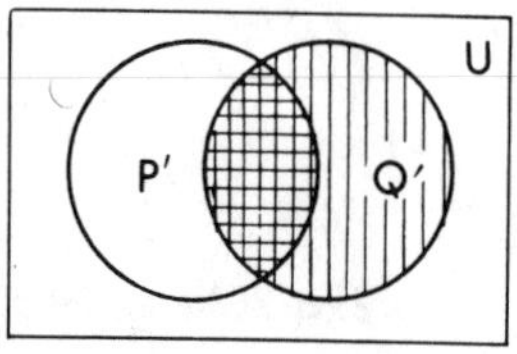

$[(\overline{P' \cap Q'}) \cap P'] \subset Q'$ so that $[(p \to q) \wedge p] \Rightarrow q$

Figure 2.26

Exercises 2.9

1. The following statements have the statement form $p \rightarrow q$. Rewrite them so that they will have the statement form $\overline{p \wedge \overline{q}}$.

a) If fire destroys London, then alligators have no teeth.

b) If Lincoln was famous, then Babe Ruth played baseball.

c) If spiders are creepy, then roaches are ugly.

d) If the circus comes to town within the next two weeks, then I shall go to a performance.

2. Rewrite the following statements in the form, "If . . . , then. . . ."

a) To get to my math class on time, I must leave my home fifteen minutes before class time.

b) To succeed, one must take risks.

c) Monkey see, monkey do.

d) I will play golf this afternoon if it is not raining.

e) Infection in the body causes the white corpuscle count to increase.

f) Those that score above 90% on the final exam will receive an A.

g) You can legally drive a car provided you are at least sixteen years of age and pass a driver's test.

h) A child will have a long and prosperous life only if he obeys his parents.

3. Show that the following is a tautology.

$$[(p \wedge \overline{q}) \rightarrow (r \wedge \overline{r})] \rightarrow (p \rightarrow q)$$

4. True or False

a) All implications are conditional statements.

b) All tautologies are conditional statements.

c) All logically true statements are tautologies, and all tautologies are implications.

d) All conditional statements are implications.

e) All implications are logically true.

f) If $p \rightarrow q$ is true, then $q \rightarrow p$ is true.

g) There are tautologies that are not conditional statement forms.

h) There are conditional statement forms that are not logically true.

i) There are tautologies that are not implications.
j) All implications are tautologies.

5. Construct truth tables for the following.

a) $\overline{p} \to \overline{q}$
b) $q \to p$
c) $p \to q$
d) $\overline{q \to p}$
e) $(p \land q) \to p$
f) $(p \lor q) \to q$
g) $p \to (q \lor p)$
h) $q \to (p \to \overline{p})$
i) $(\overline{p \lor q}) \to \overline{p}$
j) $(p \land q) \to (\overline{q} \land \overline{p})$
k) $(p \lor q) \to (\overline{p} \land \overline{q})$
l) $(\overline{q} \to \overline{p}) \land (p \to q)$
m) $[p \to (q \land r)] \to (p \lor q)$
n) $(\overline{p \lor r}) \to [q \land (\overline{r} \to p)]$
o) $[(q \to r) \to \overline{p}] \to r$
p) $[(q \land p) \to (q \lor p)] \land (r \to q)$

6. Let P be "Students are radicals," Q be "Radicals are incompetent," R be "Colleges are important," and S be "Incompetent people are students." Write each of the following in words.

a) $R \to (P \land Q)$
b) $\overline{S} \lor (\overline{R} \to P)$
c) $[S \land (P \land Q)] \Rightarrow S$
d) $(\overline{P \lor S}) \to [S \land (R \lor \overline{P})]$

7. Write the statement form of each of the following sentences.

a) R. D. is an excellent fisherman, but he has trouble if the fish aren't biting.
b) Bill takes out the trash whenever Margaret nags him about it, and he does not take out the trash if she does not nag him.
c) Seeding clouds will cause rain only if the atmospheric conditions are just right.
d) Johnny Appleseed did not exist, or if he did exist he was a hermit.
e) If toothpaste did not exist or submarines could fly, then man would have three eyes and toothpaste would exist.

8. Show by using Venn diagrams that the subset relationship holds for each of the following implications.

a) $q \Longrightarrow (q \vee \overline{q})$

b) $[p \wedge (p \rightarrow q)] \Longrightarrow q$

c) $(p \wedge q) \Longrightarrow (p \rightarrow q)$

d) $[p \wedge (q \wedge r)] \Longrightarrow p$

e) $[(q \wedge r) \wedge (\overline{p \wedge r})] \Longrightarrow [r \wedge (\overline{p \wedge q})]$

f) $[r \wedge (\overline{\overline{p} \rightarrow q})] \Longrightarrow (\overline{r \rightarrow p})$

Section 10
Biconditionals

The last connective we will consider in our study of logic will be defined in terms of the conditional. We call this connective the *biconditional*, symbolized "$\leftrightarrow$" and read "if and only if." For p and q statement variables, $p \leftrightarrow q$ will be the conjunction of the conditional statement form $p \rightarrow q$ with the conditional statement form $q \rightarrow p$. The truth table for $p \leftrightarrow q$ is given in Figure 2.27.

p	q	$p \rightarrow q$	$q \rightarrow p$	$(p \rightarrow q) \wedge (q \rightarrow p)$	$p \leftrightarrow q$
T	T	T	T	T	T
T	F	F	T	F	F
F	T	T	F	F	F
F	F	T	T	T	T

Figure 2.27

If we examine the biconditional $p \leftrightarrow q$, we note that $p \leftrightarrow q$ is true only when p and q have the same truth values. Otherwise, it is false. This fact is illustrated in the Venn diagram, which is associated with the statement form $p \leftrightarrow q$ (Figure 2.28). Let P′ and Q′, subsets of U, be truth sets for statement variables p and q. Note that we made the Venn diagram for the expression $(\overline{P' \cap \overline{Q}'}) \cap (\overline{Q' \cap \overline{P}'})$ with the appropriate shaded area for the statement form $p \leftrightarrow q$.

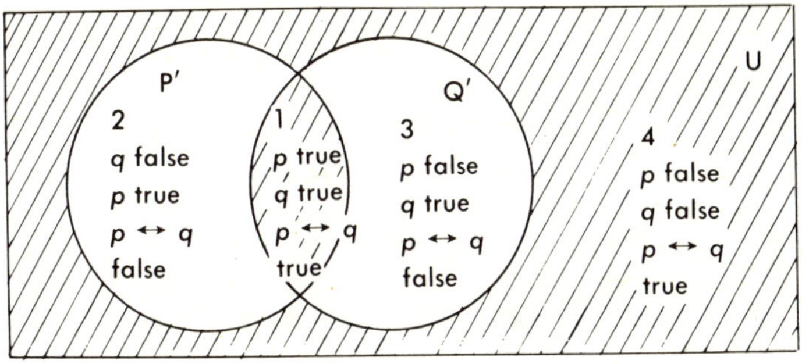

Shaded area represents the truth set for statement form $p \leftrightarrow q$.

Figure 2.28

	p	q	$p \leftrightarrow q$
1	T	T	T
2	T	F	F
3	F	T	F
4	F	F	T

Figure 2.29

We use numerals to associate the Venn diagram in Figure 2.28 with the truth table in Figure 2.29.

Let p and q be related in such a way that the statement form $p \leftrightarrow q$ is a logical truth (tautology). We then say that p is *equivalent* to q and write $p \Leftrightarrow q$. If we have $p \Leftrightarrow q$, the

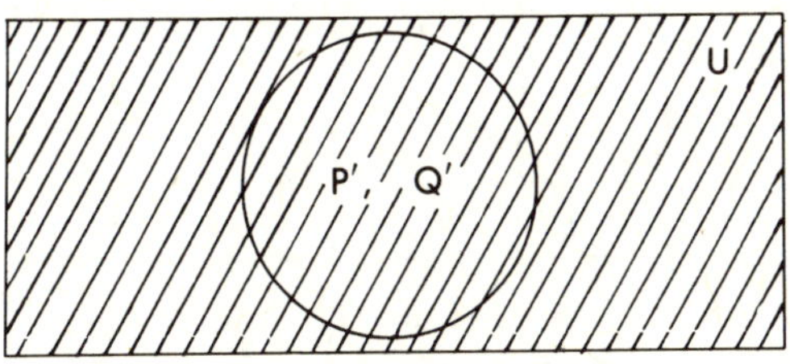

Venn diagram for $p \Leftrightarrow q$. $P' = Q'$

Figure 2.30

truth set for $p \leftrightarrow q$ is U. Hence, $P' \subset Q'$ and $Q' \subset P'$. But this means that $P' = Q'$.

The student should keep in mind that the symbol "$\leftrightarrow$" (read "if and only if) is used to compound two statements, whereas the symbol "$\Leftrightarrow$" (read "is equivalent to") not only compounds two statements but indicates a relation of equivalence between the two statements.

Consider the statement form $(\overline{p \wedge q}) \leftrightarrow (\bar{p} \vee \bar{q})$. In Figure 2.31, we have the truth table for $(\overline{p \wedge q}) \leftrightarrow (\bar{p} \vee \bar{q})$.

p	q	$\bar{p}$	$\bar{q}$	$p \wedge q$	$\overline{p \wedge q}$	$\bar{p} \vee \bar{q}$	$(\overline{p \wedge q}) \leftrightarrow (\bar{p} \vee \bar{q})$
T	T	F	F	T	F	F	T
T	F	F	T	F	T	T	T
F	T	T	F	F	T	T	T
F	F	T	T	F	T	T	T

Figure 2.31

Since $(\overline{p \wedge q}) \leftrightarrow (\bar{p} \vee \bar{q})$ is a tautology, we can write $(\overline{p \wedge q}) \Leftrightarrow (\bar{p} \vee \bar{q})$ and say that $\overline{p \wedge q}$ and $\bar{p} \vee \bar{q}$ are equivalent statement forms. Notice that the truth values for $\overline{p \wedge q}$ and $\bar{p} \vee \bar{q}$ are the same, as is indicated in Figure 2.31. In a less formal manner, we may say statement forms are equivalent if they have the "same truth tables."

The Venn diagrams in Figures 2.32 and 2.33 illustrate that the truth sets for the statement forms $(\overline{p \wedge q})$ and

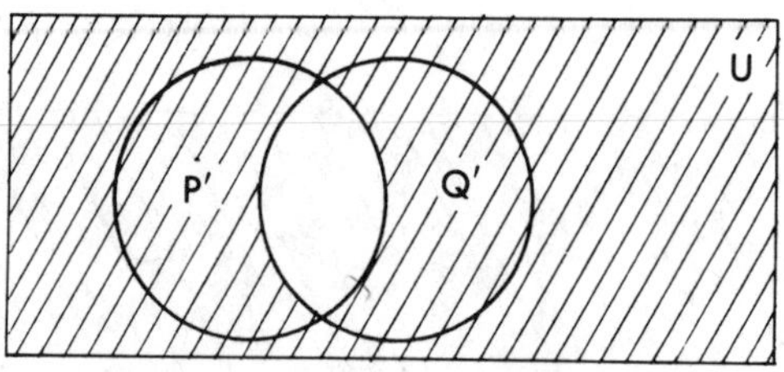

Shaded area $\overline{P' \cap Q'}$ represents truth set for $\overline{p \wedge q}$.

Figure 2.32

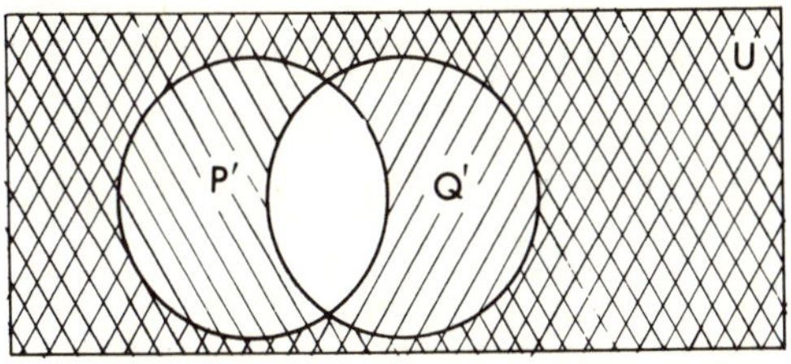

Shaded area $\overline{P'} \cup \overline{Q'}$
represents truth set for
$\bar{p} \vee \bar{q}$.

Figure 2.33

$(\bar{p} \vee \bar{q})$ are equal when P′ and Q′, subsets of U, are truth sets for p and q, respectively.

One of the important properties of equivalent statement forms is that a statement form can be substituted for any of its equivalents. This will be particularly important to remember when we get to the next chapter.

Section 11
Variations on the Conditional

Consider the conditional statement, "If I study mathematics, then I enjoy mathematics." Let P be "I study mathematics," and let Q be "I enjoy mathematics." In addition to the conditional statement $P \rightarrow Q$, there are three related conditional statements of interest. Variations on the conditional state-

Name	Example	Read	Symbolized	Instance of
Conditional statement	If I study mathemathics, then I enjoy mathematics	If P, then Q	$P \rightarrow Q$	$p \rightarrow q$
Converse statement	If I enjoy mathematics, then I study mathematics	If Q, then P	$Q \rightarrow P$	$q \rightarrow p$
Inverse statement	If I do not study mathematics, then I do not enjoy mathematics	If not P, then not Q	$\bar{P} \rightarrow \bar{Q}$	$\bar{p} \rightarrow \bar{q}$
Contrapositive statement	If I do not enjoy mathematics, then I do not study mathematics.	If not Q, then not P	$\bar{Q} \rightarrow \bar{P}$	$\bar{q} \rightarrow \bar{p}$

Figure 2.34

ment form $p \to q$ occur so often in mathematics that we give them names, as indicated in Figure 2.34.

The truth tables for statement forms $p \to q$ and $\overline{q} \to \overline{p}$ and the truth table for $(p \to q) \leftrightarrow (\overline{q} \to \overline{p})$ are given in Figure 2.35. Note that $(p \to q) \leftrightarrow (\overline{q} \to \overline{p})$ is a tautology. Hence, a conditional statement form and its contrapositive statement form are equivalent, i.e., $(p \to q) \Leftrightarrow (\overline{q} \to \overline{p})$.

p	q	$\bar{p}$	$\bar{q}$	$p \to q$	$\bar{q} \to \bar{p}$	$(p \to q) \leftrightarrow (\bar{q} \to \bar{p})$
T	T	F	F	T	T	T
T	F	F	T	F	F	T
F	T	T	F	T	T	T
F	F	T	T	T	T	T

Figure 2.35

Figure 2.36 shows the truth tables for statement forms $(q \to p)$ and $(\overline{p} \to \overline{q})$, and the truth table for $(q \to p) \leftrightarrow (\overline{p} \to \overline{q})$. It is clear that $q \to p$ is equivalent to $\overline{p} \to \overline{q}$.

The truth table in Figure 2.37 shows that the statement form $(p \to q) \leftrightarrow (q \to p)$ is not a tautology and, hence,

p	q	$\bar{p}$	$\bar{q}$	$q \to p$	$\bar{p} \to \bar{q}$	$(q \to p) \leftrightarrow (\bar{p} \to \bar{q})$
T	T	F	F	T	T	T
T	F	F	T	T	T	T
F	T	T	F	F	F	T
F	F	T	T	T	T	T

Figure 2.36

p	q	$p \to q$	$q \to p$	$(p \to q) \leftrightarrow (q \to p)$
T	T	T	T	T
T	F	F	T	F
F	T	T	F	F
F	F	T	T	T

Figure 2.37

that p $\rightarrow q$ and $q \rightarrow p$ are not equivalent. Let P be "I live in Phoenix," and let Q be "I live in Arizona." It is true that P → Q (If I live in Phoenix, then I live in Arizona), but false that Q → P (If I live in Arizona, then I live in Phoenix).

Perhaps one of the most common errors in reasoning is to suppose that because a conditional statement is true, the converse statement is also true. The authors warn you of this pitfall so that your reasoning may be correct and your days in mathematics courses be enjoyable.

Section 12
Discussion

We gave the phrase "p only if q" as another way of phrasing "If p, then q." Now, "p only if q" means "If $\overline{q}$, then $\overline{p}$." But $\overline{q} \rightarrow \overline{p}$ is equivalent to $p \rightarrow q$. Thus, it is reasonable to say the phrase "p only if q" means the same as the phrase "If p, then q."

Sometimes we need to negate a conditional statement form. To do this, we note that $(\overline{p \rightarrow q}) \Leftrightarrow (p \wedge \overline{q})$. (You should check this.) Hence $p \wedge \overline{q}$ is a convenient way to denote the negation of a conditional. This form for a negated conditional should not be surprising, since we defined $p \rightarrow q$ to mean $\overline{p \wedge \overline{q}}$. Suppose we let P be "I get a job" and let Q be "I will buy a car." Then $\mathrm{P} \rightarrow \mathrm{Q}$ symbolizes, "If I get a job, then I will buy a car." Consider $\overline{\mathrm{P} \rightarrow \mathrm{Q}}$. This symbolizes, "It is not the case that if I get a job, then I will buy a car." Since $(\overline{\mathrm{P} \rightarrow \mathrm{Q}}) \Leftrightarrow (\mathrm{P} \wedge \overline{\mathrm{Q}})$, another way of expressing the negation of $\mathrm{P} \rightarrow \mathrm{Q}$ is to say, "I got a job, but I am not going to buy a car."

In this and the preceding chapter, we state many of our definitions in the form "p if and only if q." We point out that another way of reading "p if and only if q" is "p if q and p only if q," which is equivalently expressed as "If q, then p, and if p, then q." Hence, "p if and only if q" is the conjunction of a conditional and its converse. In a definition we try to formulate the statement in such a manner that it includes all desired cases and excludes all undesired cases. For definitions this means that "p if and only if q" is logically true. This discussion should heighten your understanding of the phrase "if and only if," as well as give a rationale for the use of the phrase in our definitions.

Figure 2.38 shows a summary of the connectives we have presented. It is important to remember that the connectives are defined by the truth tables given for each. The symbol "$\Rightarrow$," read "implies," can replace the symbol "$\rightarrow$" only when the conditional is a logical truth. Also, "$\Leftrightarrow$," read "is equivalent to," can be used in place of "$\leftrightarrow$" only when the biconditional is a logical truth.

Name	Symbol	Read	Defined by truth table in:
Negation	—	not	Fig. 2.4
Conjunction	$\wedge$	and	Fig. 2.6
Disjunction	$\vee$	or	Fig. 2.8
Conditional	$\rightarrow$	If . . . , then . . .	Fig. 2.20
Biconditional	$\leftrightarrow$	. . . if and only if . . .	Fig. 2.27

Figure 2.38

Exercises 2.10, 11, 12

1. True or False

a) If $p \leftrightarrow q$ is logically true, then $q \leftrightarrow p$ is logically true.

b) p and q are said to be equivalent if $p \leftrightarrow q$ is a tautology.

c) $p \leftrightarrow q$ is true only when p and q have the same truth values.

d) The set relationship associated with a biconditional statement is that of set equality.

e) The set relationship associated with an equivalent statement is that of set equality.

f) If $p \Leftrightarrow q$, then the truth table for p must be the same as the truth table for q.

g) Any statement form may be substituted for any of its logical equivalents.

h) $p \rightarrow q$ is equivalent to $\bar{q} \rightarrow \bar{p}$.
i) $q \rightarrow p$ is equivalent to $\bar{q} \rightarrow \bar{p}$.
j) $p \rightarrow q$ is equivalent to $q \rightarrow p$.
k) A conditional statement and its contrapositive are equivalent.
l) The inverse and converse of a conditional statement are equivalent.
m) The converse of a conditional statement form will have the exact opposite truth values to those of the conditional statement form itself.

2. Show that $(\overline{p \wedge \bar{q}}) \Leftrightarrow (\bar{p} \vee q)$ and, hence, $\bar{p} \vee q$ also depicts $p \rightarrow q$.

3. Let P be "Shakespeare deserved his fame" and let Q be "Shakespeare's works should be read by everyone." Write the following in sentence form.

a) $P \rightarrow Q$
b) The converse of $P \rightarrow Q$.
c) The inverse of $P \rightarrow Q$.
d) The contrapositive of $P \rightarrow Q$.

4. Let $\overline{P}$ be "Helen cannot find her glasses" and let $\overline{Q}$ be "Helen cannot drive." Write the following statements in sentence form.

a) $P \rightarrow Q$
b) The converse of $P \rightarrow Q$.
c) The inverse of $P \rightarrow Q$.
d) The contrapositive of $P \rightarrow Q$.

5. Show that the following are equivalent statement forms by
1) constructing the truth table, and
2) drawing the associated Venn diagram.

a) $[(p \wedge q) \wedge r] \Leftrightarrow [p \wedge (r \wedge q)]$
b) $[p \wedge (\overline{q \vee r})] \Leftrightarrow [(p \vee q) \wedge (\bar{q} \wedge \bar{r})]$
c) $(\overline{\bar{p} \vee q}) \Leftrightarrow (\bar{q} \wedge \bar{p})$
d) $(\overline{p \rightarrow q}) \Leftrightarrow (p \wedge \bar{q})$

6. Give the negation of the following statements in sentence form.

a) If the hatchet is dull, you will be wasting your efforts chopping the wood.

b) Visitors can make themselves at home only if they know where the kitchen is.

c) Carl should not put his elbows on the table or start eating his dessert before the hostess sits down if he wants to exemplify good manners.

d) When the sun goes down and the wind begins to blow, people at Candlestick Park either get extremely cold or they go home.

7. Simplify the following statement forms by substituting equivalent forms.

a) $(\overline{p \wedge \bar{q}}) \vee (\bar{q} \rightarrow \bar{p})$

b) $(\overline{\bar{p} \wedge q}) \wedge (\bar{p} \rightarrow \bar{q})$

c) $\overline{\bar{r} \wedge (\overline{q \wedge \bar{p}}}) $

d) $(\bar{p} \vee \bar{q}) \wedge (\bar{q} \rightarrow p)$

e) $[\overline{(\bar{p} \rightarrow \bar{q}) \wedge (\overline{p \vee q}})] \wedge [\overline{(\bar{p} \wedge \bar{q}) \wedge (\overline{p \wedge q}})]$

Section 13

Sets and Statements

If we think of a universal set U as a set of logical possibilities and think of subsets of U as being truth sets for statement variables, it is clear that for any expression about these truth sets there is an associated statement form. It is also clear that for each statement form there is an associated expression about sets.

The table in Figure 2.39 should help the student keep in mind the appropriate associations regarding sets and statement forms.

Sets	Statement forms
P'	p
$P' \cap Q'$	$p \wedge q$
$P' \cup Q'$	$p \vee q$
$P' \subset Q'$	$p \Longrightarrow q$
$P' = Q'$	$p \Longleftrightarrow q$
$\bar{P}'$	$\bar{p}$
$\bar{P}' \cap \bar{Q}'$	$\bar{p} \wedge \bar{q}$
$\bar{P}' \cup \bar{Q}'$	$\bar{p} \vee \bar{q}$

Let p and q be statement variables and let P' and Q', subsets of U, be the truth sets of p and q, respectively.

Figure 2.39

Section 14
Some Important Tautologies

To determine the truth value of a simple statement, we check experience or previously established knowledge. In general, to determine the truth value of a compound statement, we must not only know the truth values of the component statements but we must consider the construction of the compound statement (form of the statement).

If a statement is an instance of a tautologous statement form, we know that the statement is true, regardless of the truth values of its component statements. Statement forms that are tautologous (logically true) are the very heart of logic. Many tautologous statement forms are so frequently referred to in logic that they have been given names. Some of these important statement forms are listed below. The listing is divided into two categories. First, we list some tautologous conditional statement forms; second, we list some tautologous biconditional statement forms. If you are wise, you will make sure that, in fact, all of the following statement forms are tautologies. (We might be pulling a fast one.)

Tautologous Conditional Statement Forms

1. $[(p \rightarrow q) \wedge p] \Rightarrow q$ Law of detachment (L.D.)
2. $[(p \rightarrow q) \wedge \overline{q}] \Rightarrow \overline{p}$ Indirect argument (I.A.)
3. $[(p \rightarrow q) \wedge (q \rightarrow r)] \Rightarrow (p \rightarrow r)$ Law of syllogism (L.S.)

4. $[(p \vee q) \wedge \bar{p}] \Rightarrow q$ or $[(p \vee q) \wedge \bar{q}] \Rightarrow p$ Alternative argument (A.A.)

5. $[(p \rightarrow q) \wedge (r \rightarrow s) \wedge (p \vee r)] \Rightarrow (q \vee s)$ Constructive dilemma (C.D.)

6. $(p \rightarrow q) \Rightarrow [p \rightarrow (p \wedge q)]$ Absorption (Abs.)

7. $[(p \rightarrow q) \wedge (r \rightarrow s) \wedge (\bar{q} \vee \bar{s})] \Rightarrow (\bar{p} \vee \bar{r})$ Destructive dilemma (D.D.)

8. $(p \wedge q) \Rightarrow p$ or $(p \wedge q) \Rightarrow q$ Simplification (Simp.)

9. $p \Rightarrow (p \vee q)$ Addition (Add.)

10. $p, q \Rightarrow (p \wedge q)$ Conjunction (Conj.)

Equivalent Statement Forms

1. De Morgan's Laws (De M.)

$(\overline{p \wedge q}) \Leftrightarrow (\bar{p} \vee \bar{q})$

$(\overline{p \vee q}) \Leftrightarrow (\bar{p} \wedge \bar{q})$

2. Commutation (Comm.)

$(p \vee q) \Leftrightarrow (q \vee p)$

$(p \wedge q) \Leftrightarrow (q \wedge p)$

3. Association (Assoc.)

$[p \wedge (q \wedge r)] \Leftrightarrow [(p \wedge q) \wedge r]$

$[p \vee (q \vee r)] \Leftrightarrow [(p \vee q) \vee r]$

4. Distributivity (Dist.)

$[p \wedge (q \vee r)] \Leftrightarrow [(p \wedge q) \vee (p \wedge r)]$

$[p \vee (q \wedge r)] \Leftrightarrow [(p \vee q) \wedge (p \vee r)]$

5. Double negation (D.N.)

$p \Leftrightarrow \bar{\bar{p}}$

6. Contrapositive (Contra.)

$(p \rightarrow q) \Leftrightarrow (\bar{q} \rightarrow \bar{p})$

7. Conditional (Cond.)

$(p \rightarrow q) \Leftrightarrow (\bar{p} \vee q)$

$(p \vee q) \Leftrightarrow (p \rightarrow q)$

8. Equivalence (Equiv.)

$(p \leftrightarrow q) \Leftrightarrow [(p \rightarrow q) \wedge (q \rightarrow p)]$

$(p \leftrightarrow q) \Leftrightarrow [(p \wedge q) \vee (\bar{p} \wedge \bar{q})]$

9. Exportation (Exp.)

$[(p \wedge q) \rightarrow r] \Leftrightarrow [p \rightarrow (q \rightarrow r)]$

10. Reduction (Red.) $p \Leftrightarrow (p \vee p)$ and $p \Leftrightarrow (p \wedge p)$

Notice that the associative property of "$\wedge$" and "$\vee$" means that we can write $p \wedge q \wedge r$ and $p \vee q \vee r$ without ambiguity.

Exercises 2.13, 14

1. Let U be the set of all logical possibilities for p, q, and r. Let P′, Q′, and R′ be truth sets for p, q, and r, respectively. For each of the set expressions below, construct the Venn diagram and give the associated statement form.

a) $P' - Q'$
b) $P' \cap (Q' - P')$
c) $(Q' \cap \bar{P}') \subset (Q' \cup P')$
d) $[Q' - (P' \cap Q')] \subset Q'$
e) $(\bar{P}' \cap \bar{Q}') \cap \bar{R}' = [\overline{(P' \cup R') \cup Q'}]$
f) $[(R' \cup P') \cap (P' - R')] \subset [(P' \cup R') - (P' \cap R')]$
g) $R' - [\bar{Q}' \cap (P' - R')]$
h) $[Q' \cap (\bar{P}' \cup \bar{R}')] - [P' - (Q' \cap P')]$
i) $[(P' \cup R') - P'] - Q'$

2. Let U be the set of all logical possibilities for p, q, and r. Let P′, Q′, and R′ be truth sets for p, q, and r, respectively. For each of the statement forms below give the associated set expression.

a) $[(p \rightarrow q) \wedge p] \Rightarrow q$
b) $(q \wedge \bar{p}) \Leftrightarrow [(p \vee q) \wedge \bar{p}]$
c) $[(p \rightarrow q) \wedge (r \rightarrow q) \wedge \bar{r}] \Rightarrow p$
d) $p \Rightarrow (p \vee q)$
e) $(\overline{p \wedge \bar{q}}) \Leftrightarrow (\bar{p} \vee q)$

3. Show that the 10 conditional statement forms given in Section 14 are tautologies by constructing the truth tables for each one.

4. Show that the 10 biconditional statement forms given in Section 14 are tautologies by constructing the truth tables for each one.

Section 15
Concluding Remarks

In Figure 2.40, we show the highlights of our development. The diagram is provided as a means of helping you stabilize your "canoe of understanding" in our "stream of development."

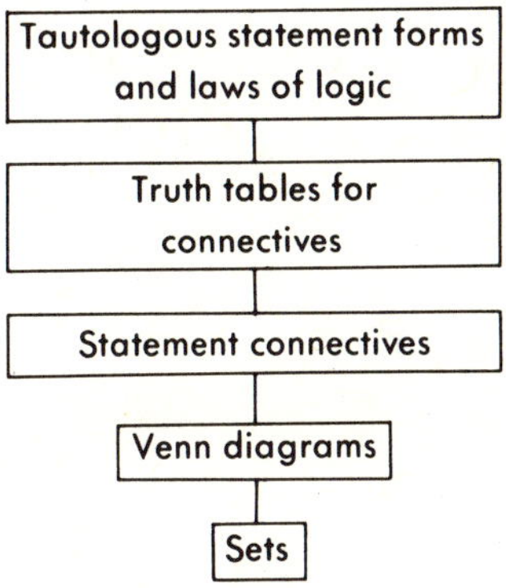

Figure 2.40

This chapter illustrates the beauty of how mathematical concepts cast in one setting melt into mathematical concepts cast in another setting. The interrelatedness of mathematical ideas produces a charm and continuity that woos the mind to probe the innermost mysteries of mathematics.

You undoubtedly appreciate the delightful aspects of what you have been studying in these chapters, but without the perspective of a mature mathematical eye, you may be inclined to say, "This is all very interesting, but so what?" The authors do not claim that studying mathematics and, in particular, studying logic will cause you suddenly to blossom into a mental giant. However, we do feel that a knowledge of these foundational ideas will be useful not only in further mathematical studies but also in everyday experience. Also the aesthetic nature of the abstract edifice of thought we are building will add another jewel in your repertoire of things to enjoy, such as art, music, and poetry.

References
Chapter Two

Copi, Irving M., *Introduction to Logic,* 3rd ed. (London: The Macmillan Company, 1968).

Kemeny, John G.; Snell, J. Laurie; and Thompson, Gerald L., *Introduction to Finite Mathematics* (Englewood Cliffs, N.J.: Prentice-Hall, Inc., 1957) pp. 63–66.

Schagrin, Morton L., *The Language of Logic—A Programmed Text* (New York: Random House, 1968).

CHAPTER THREE

Deductive Reasoning

Section 1

Some Ways to Show Arguments Valid

The process of reasoning by which one judges a statement in the light of accepted statements is called *deductive reasoning*. In attempting to focus on deductive reasoning strategies in this chapter, the authors do not intend to minimize other types of thinking nor suggest that different types of thinking can be clearly separated. The deductive reasoning process is used by each of us every day. In mathematics, perhaps as in no other area of endeavor, the deductive reasoning process is exemplified in its purest form. The concepts, principles, terminology, and notation presented in Chapter 2 provide us with an excellent foundation for a study of the deductive reasoning process as it relates to a logical system.

Chapter 3 will provide you with a model exemplifying the deductive reasoning process. This chapter will show the importance of linguistic considerations in deductive reasoning and, hopefully, will help you appreciate the deductive reasoning process and perhaps help you employ it more precisely and perceptively in your daily experience. The remaining chapters rely heavily on understanding the deductive process and the principles of symbolic logic.

We will frequently employ the term "argument" in the following discussion. In logic, the term "argument" is not used to indicate a fight or dispute. For us, an *argument* will be a sequence of two or more statements, at least one of which is to be established on the basis of the others. We call the statement that is to be established the *conclusion*. The other supporting statements are called *premises*. Consider the following sequence of statements.

> If I go to the party, then I will be out late.
> I am going to the party.
> Hence, I will be out late.

This sequence of statements is an argument. Letting P be "I am going to the party" and letting Q be "I will be out late," we symbolize this argument as follows.

$$\begin{array}{l} P \rightarrow Q \\ P \\ \hline \text{Hence, } Q \end{array} \qquad \text{or} \qquad \begin{array}{l} P \rightarrow Q \\ P \\ \hline \therefore Q \end{array}$$

The three dots in a triangle mean "hence," "therefore," or some equivalent expression. The statements above the line are called premises, and the statement below the line is called the conclusion.

The forms of arguments are more useful for us to study than the arguments themselves, as was the case with statements and the forms of statements. An *argument form* is an arrangement of statement variables so that when statements replace the statement variables an argument results. When the statement variables in an argument form are replaced with specific statements, we have an argument that is an *instance* of that argument form. For example,

$$\begin{array}{l} p \rightarrow q \\ p \\ \hline \therefore q \end{array}$$

is an array of statement variables that is an argument form. Notice that

$$\begin{array}{l} P \rightarrow Q \\ P \\ \hline \therefore Q \end{array}$$

is an argument (for P and Q as designated in the previous

paragraph) that is an instance of the argument form

$$\begin{array}{l} p \rightarrow q \\ \underline{p \qquad} \\ \therefore q \end{array}$$

In dealing with arguments, we are primarily concerned with whether or not a stated conclusion is indeed established by the given premises. Conditional statement forms play a central role in determining whether or not a stated conclusion is established by given premises. Consider the argument form of which a given argument is an instance. This argument form is related to a conditional statement form in the following way: If you conjunct all premises of an argument within a grouping symbol, replace the line in the argument form with the connective $\rightarrow$, and put the conclusion to the right of $\rightarrow$, you have a conditional statement form. Notice that the hypothesis of the conditional statement form is the conjunction of the premises of the argument form and that the conclusion of the conditional statement form is the conclusion of the argument form. In this manner, one can always associate a conditional statement form with a given argument form. Also, notice that for any given conditional statement form, one can determine an associated argument form. This is done by writing the individual conjuncts that are to the left of $\rightarrow$ in column fashion, replacing $\rightarrow$ by a line, and writing what is to the right of $\rightarrow$ below the line. We have noted that

$$\begin{array}{l} \mathrm{P} \rightarrow \mathrm{Q} \\ \underline{\mathrm{P} \qquad} \\ \therefore \mathrm{Q} \end{array}$$

is an argument that is an instance of the argument form

$$\begin{array}{l} p \rightarrow q \\ \underline{p \qquad} \\ \therefore q \end{array}$$

The conditional statement form associated with this argument form is $[(p \rightarrow q) \wedge p] \rightarrow q$.

Tautologous conditional statement forms will be the key to helping us decide if a stated conclusion is established by the given premises. Consider again the following argument.

> If I go to the party, then I will be out late.
> I am going to the party.
> Hence, I will be out late.

We would all agree that the premises of this argument support the conclusion. Hence, we would consider this argument to be an example of correct reasoning. Notice that the argument

$$\begin{array}{l} P \rightarrow Q \\ \underline{P \qquad\quad} \\ \therefore\ Q \end{array}$$

is an instance of the argument form

$$\begin{array}{l} p \rightarrow q \\ \underline{p \qquad\quad} \\ \therefore\ q, \end{array}$$

whose associated statement form $[(p \rightarrow q) \wedge p] \rightarrow q$ is a tautology. With the statements P and Q, as designated in the argument

$$\begin{array}{l} P \rightarrow Q \\ \underline{P \qquad\quad} \\ \therefore\ Q, \end{array}$$

we can form the conditional statement $[(P \rightarrow Q) \wedge P] \rightarrow Q$, which is an instance of the tautologous conditional statement form $[(p \rightarrow q) \wedge p] \rightarrow q$. This conditional statement is true since it is an instance of a tautology. We note that when the hypothesis of a conditional statement is true and the conditional statement itself is true, the conclusion of the conditional statement must be true. Hence, the conclusion Q is true when the premises $P \rightarrow Q$ and P are true. That is, the premises support the conclusion. This is the way conditional statement forms help us decide if an argument is, in fact, correct. If an argument is an instance of an argument form whose associated conditional statement form is tautologous, we know that, if the premises of the argument are true, the conclusion is true. The following definitions embody the ideas presented in the previous discussion.

Definition: An argument form is called a *valid argument form* if and only if the associated conditional statement form is a tautology.

Definition: An argument is called a *valid argument* if and only if it is an instance of a valid argument form.

From our previous work, we know that $[(p \rightarrow q) \wedge p] \rightarrow q$ is a tautology. Hence,

$$\begin{array}{l} p \rightarrow q \\ \underline{p \qquad\;} \\ \therefore q \end{array}$$

is a valid argument form and

$$\begin{array}{l} \mathrm{P} \rightarrow \mathrm{Q} \\ \underline{\mathrm{P} \qquad\;} \\ \therefore \mathrm{Q} \end{array}$$

is a valid argument since it is an instance of a valid argument form.

There are some valid argument forms which appear quite frequently. These valid argument forms correspond to the tautologous conditional statement forms given in Section 14 of Chapter 2 and are indicated in Figure 3.1.

Consider the following argument.

Either the water is on or the furnace is on.
The furnace is not on.

∴ The water is on.

This argument can be symbolized as

$$\begin{array}{l} \mathrm{P} \vee \mathrm{Q} \\ \underline{\overline{\mathrm{Q}} \qquad\;} \\ \therefore \mathrm{P} \end{array},$$

when P and Q represent the statements in the argument.

Valid Argument Forms

(1) Law of detachment

$$\begin{array}{l} p \rightarrow q \\ \underline{p \qquad} \\ \therefore q \end{array}$$

(2) Indirect argument

$$\begin{array}{l} p \rightarrow q \\ \underline{\overline{q} \qquad} \\ \therefore \overline{p} \end{array}$$

(3) Law of syllogism

$$\begin{array}{l} p \rightarrow q \\ \underline{q \rightarrow r} \\ \therefore p \rightarrow r \end{array}$$

(4) Alternative argument

$$\begin{array}{l} p \vee q \\ \underline{\overline{p} \qquad} \\ \therefore q \end{array} \quad \text{or} \quad \begin{array}{l} p \vee q \\ \underline{\overline{q} \qquad} \\ \therefore p \end{array}$$

(5) Constructive dilemma

$$\begin{array}{l} p \rightarrow q \\ r \rightarrow s \\ \underline{p \vee r} \\ \therefore q \vee s \end{array}$$

(6) Destructive dilemma

$$\begin{array}{l} p \rightarrow q \\ r \rightarrow s \\ \underline{\overline{q} \vee \overline{s}} \\ \therefore \overline{p} \vee \overline{r} \end{array}$$

(7) Absorption

$$\begin{array}{l} \underline{p \rightarrow q \qquad} \\ \therefore p \rightarrow (p \wedge q) \end{array}$$

(8) Simplification

$$\begin{array}{l} \underline{p \wedge q} \\ \therefore p \end{array} \quad \text{or} \quad \begin{array}{l} \underline{p \wedge q} \\ \therefore q \end{array}$$

(9) Addition

$$\begin{array}{l} \underline{p \qquad} \\ \therefore p \vee q \end{array}$$

(10) Conjunction

$$\begin{array}{l} p \\ \underline{q \qquad} \\ \therefore p \wedge q \end{array}$$

Figure 3.1

Now

$$\begin{array}{l} p \vee q \\ \underline{\overline{q} \qquad} \\ \therefore p \end{array}$$

is the argument form of which the given argument is an instance. Since

$$\begin{array}{l} P \vee Q \\ \underline{\overline{Q} \qquad} \\ \therefore P \end{array}$$

is an instance of valid argument form Number 4, Figure 3.1, the argument is valid.

The following argument is also valid.

If I wear green socks, then cows are quadrupeds.
If cows are quadrupeds, then men have landed on the moon.

∴ If I wear green socks, then men have landed on the moon.

Symbolically, this argument is

$$\begin{array}{l} P \rightarrow Q \\ \underline{Q \rightarrow R} \\ \therefore\ P \rightarrow R \end{array},$$

and it is an instance of argument form Number 3, Figure 3.1. Note that while the above argument is valid, it does not express anything from real life discourse. In mathematical reasoning, the premises of an argument are not required to express "true-to-life" situations. A valid argument simply states what logically follows if one accepts the premises to be true.

There are several methods for determining whether or not an argument is valid. Of these, we will discuss two. One of these methods is by determining the argument form associated with the argument, converting the argument form to a conditional statement form, and then building a truth table for the statement form. If the statement form is tautologous, then the argument is valid, since it is an instance of a valid argument form. Of course, if an argument is an instance of one of the ten listed argument forms (Figure 3.1), then we know immediately that the argument is valid.

Consider the following argument: If the insulation is poor, then the pipes will freeze. If the insulation protects the pipes in below-freezing weather, then the insulation is expensive. Either the insulation is poor or the insulation protects the pipes in below-freezing weather. The insulation is not expensive. Therefore, the pipes will freeze.

Let the following symbols represent the indicated statements.

P: The insulation is poor.

Q: The pipes will freeze.
R: The insulation protects the pipes in below-freezing weather.
S: The insulation is expensive.

The above argument can be symbolized as:

$$\begin{array}{l} P \rightarrow Q \\ R \rightarrow S \\ P \vee R \\ \overline{S} \\ \hline \therefore Q \end{array}$$

We will show by each of two methods that this is a valid argument.

Method 1: Use of Truth Tables

The argument

$$\begin{array}{l} P \rightarrow Q \\ R \rightarrow S \\ P \vee R \\ \overline{S} \\ \hline \therefore Q \end{array}$$

has the associated argument form

$$\begin{array}{l} p \rightarrow q \\ r \rightarrow s \\ p \vee r \\ \bar{s} \\ \hline \therefore q \end{array}$$

and the corresponding conditional statement form $[(p \rightarrow q) \wedge (r \rightarrow s) \wedge (p \vee r) \wedge \bar{s}] \rightarrow q$. In Figure 3.2, the truth table for this statement form is presented. In the table H is used for $(p \rightarrow q) \wedge (r \rightarrow s) \wedge (p \vee r) \wedge \bar{s}$. Since the conditional statement form is tautologous, the argument is valid.

p	q	r	s	$p \to q$	$r \to s$	$p \vee r$	$\bar{s}$	H $(p \to q) \wedge (r \to s) \wedge (p \vee r) \wedge \bar{s}$	$H \to q$
T	T	T	T	T	T	T	F	F	T
T	T	T	F	T	F	T	T	F	T
T	T	F	T	T	T	T	F	F	T
T	T	F	F	T	T	T	T	T	T
T	F	T	T	F	T	T	F	F	T
T	F	T	F	F	F	T	T	F	T
T	F	F	T	F	T	T	F	F	T
T	F	F	F	F	T	T	T	F	T
F	T	T	T	T	T	T	F	F	T
F	T	T	F	T	F	T	T	F	T
F	T	F	T	T	T	F	F	F	T
F	T	F	F	T	T	F	T	F	T
F	F	T	T	T	T	T	F	F	T
F	F	T	F	T	F	T	T	F	T
F	F	F	T	T	T	F	F	F	T
F	F	F	F	T	T	F	T	F	T

Figure 3.2

Method 2: Use of Valid Argument Forms

A second method of showing that an argument is valid uses the following format. Make a table of two columns. The column at the left contains the symbolization of specific true statements (each line is identified by a numeral). In the column at the right, there is a numeral associated with an expression indicating by what authority the numerically corresponding statement is true. We head these volumes "Statements" and "Reasons," respectively. Thus, for the argument under discussion, we have the format given in Figure 3.3.

Since Q is true when $P \rightarrow Q$, $R \rightarrow S$, $P \vee R$, and $\bar{S}$ are true, and since the only time a conditional can be false is if $T \rightarrow F$, the conditional statement form $[(p \rightarrow q) \wedge (r \rightarrow s) \wedge (p \vee r) \wedge \bar{s}] \rightarrow q$ associated with the argument form for the argument

$$\begin{array}{l} P \rightarrow Q \\ R \rightarrow S \\ P \vee R \\ \underline{\bar{S}} \\ \therefore Q \end{array}$$

is a tautology. Hence,

Statements	Reasons
1. $P \rightarrow Q$	1. Premise
2. $R \rightarrow S$	2. Premise
3. $P \vee R$	3. Premise
4. $Q \vee S$	4. Lines 1, 2, 3, and an instance of valid argument form C.D. (Figure 3.1)
5. $\bar{S}$	5. Premise
6. Q	6. Lines 4 and 5 and an instance of valid argument form A.A. (Figure 3.1)

Figure 3.3

$$\begin{array}{l} P \to Q \\ R \to S \\ P \lor R \\ \overline{S} \\ \hline \therefore Q \end{array}$$

is an instance of a valid argument form and therefore is a valid argument.

As can be seen, Method 1 is the more lengthy and time-consuming of the two methods. Although in theory Method 1 can always be used to determine if an argument is valid, it is not very practical. In particular, if one had an argument with an associated conditional statement form using seven statement variables, the associated truth table would have 128 entries. (Not only would it be physically exhausting to construct a table of this size, but you might deplete your financial resources just buying the paper.)

Method 2 is the one we shall emphasize. It involves the construction of a sequence of statements in which each line must be either a premise of an argument or follow from preceding lines by a valid argument form. The last line of the sequence is the conclusion of the argument. Such a sequence of statements is called a *proof* of the validity of an argument. The "reasons" given for each line are not actually a part of the proof, but they serve to indicate justification for putting each line in the proof. When a conclusion of an argument is shown to be true by direct use of the stated premises, as in Method 2, the proof is said to be a *direct proof.*

A direct proof of the indicated argument

$$\begin{array}{l} (P \land Q) \to R \\ R \to T \\ \overline{S} \\ (P \land Q) \lor S \\ \hline \therefore T \end{array}$$

is shown in Figure 3.4. Notice that we have shortened our expressions in the "Reasons" column, indicating only the valid argument form used and the particular lines involved in

Statements	Reasons
1. $(P \wedge Q) \vee S$	1. Premise
2. $\overline{S}$	2. Premise
3. $P \wedge Q$	3. 1, 2, and A.A.
4. $(P \wedge Q) \rightarrow R$	4. Premise
5. R	5. 3, 4, and L.D.
6. $R \rightarrow T$	6. Premise
7. T	7. 5, 6, and L.D.

Figure 3.4

that valid argument form. It is important to observe that although lines 1, 2, and 3 of our proof do not match exactly the valid argument form A.A., we can show that, in fact, lines 1, 2, and 3 are an instance of valid argument form A.A. by substituting the statement $P \wedge Q$ for p and the statement S for q; that is, $p = P \wedge Q$ and $q = S$. Then,

$$\begin{array}{l} p \vee q \\ \underline{\overline{q} \qquad} \\ \therefore p \end{array}$$

becomes

$$\begin{array}{l} (P \wedge Q) \vee S \\ \underline{\overline{S} \qquad\qquad} \\ \therefore P \wedge Q \end{array}$$

It is important to be alert to such substitutions in working with valid argument forms, because awareness of substitutions will be necessary in proving some arguments valid.

The order in which we use the premises in Figure 3.4 is not the same as the order in which the premises are given in the argument. The order of the premises as they appear in an argument and the order of their use in a proof has no bearing on the validity of an argument.

We further point out that the proof in Figure 3.4 is not the only proof that can be given for the argument's validity. We could have used the proof in Figure 3.5.

Statements	Reasons
1. $(P \wedge Q) \to R$	1. Premise
2. $R \to T$	2. Premise
3. $(P \wedge Q) \to T$	3. 1, 2, and L.S.
4. $(P \wedge Q) \vee S$	4. Premise
5. $\overline{S}$	5. Premise
6. $P \wedge Q$	6. 4, 5, and A.A.
7. T	7. 3, 6, and L.D.

Figure 3.5

You might be able to think of another proof for the validity of the argument.

$$\begin{array}{l} (P \wedge Q) \to R \\ R \to T \\ \overline{S} \\ (P \wedge Q) \vee S \\ \hline \therefore T \end{array}$$

In general, the proof for an argument's validity is not unique.

Exercises 3.1

1. Determine the appropriate substitutions to show that the following indicated arguments are actually instances of a valid argument form. Also, state the argument form for which each of the following is an instance.

a) $$\begin{array}{l} \overline{P} \to \overline{Q} \\ \overline{P} \\ \hline \therefore \overline{Q} \end{array}$$

b) $$\begin{array}{l} (P \wedge Q \wedge S) \to (R \wedge T) \\ P \wedge Q \wedge S \\ \hline \therefore R \wedge T \end{array}$$

c) $$\begin{array}{l} P \to \overline{Q} \\ \hline \therefore P \to (P \wedge \overline{Q}) \end{array}$$

d) $$\begin{array}{l} P \\ \hline \therefore P \vee \overline{Q} \end{array}$$

e) $R \rightarrow S$
$P \vee R$
$P \rightarrow Q$
$\therefore Q \vee S$

f) $\bar{R} \vee \bar{U}$
$(P \wedge Q) \rightarrow R$
$(S \wedge T) \rightarrow U$
$\therefore (\overline{P \wedge Q}) \vee (\overline{S \wedge T})$

g) $(P \vee Q) \vee (S \vee T)$
$\overline{S \vee T}$
$\therefore P \vee Q$

h) $P \rightarrow [Q \rightarrow (S \vee T)]$
P
$\therefore Q \rightarrow (S \vee T)$

i) $P \rightarrow (Q \wedge R)$
$\overline{Q \wedge R}$
$\therefore \bar{P}$

j) $P \wedge Q$
R
$\therefore (P \wedge Q) \wedge R$

2. Show that the following indicated argument is valid by using the two methods discussed in this section.

$$P \rightarrow (Q \vee R)$$
$$P \wedge \bar{R}$$
$$\therefore Q$$

Section 2
Theorems

By a procedure analogous to associating conditional statement forms and argument forms, we can associate conditional statements and arguments. For instance, the argument

$$\begin{array}{l} P \rightarrow Q \\ R \rightarrow S \\ P \vee R \\ \overline{S} \\ \hline \therefore Q \end{array}$$

and the conditional statement $[(P \rightarrow Q) \wedge (R \rightarrow S) \wedge (P \vee R) \wedge \overline{S}] \rightarrow Q$ are associated. Notice that the conditional statement $[(P \rightarrow Q) \wedge (R \rightarrow S) \wedge (P \vee R) \wedge \overline{S}] \rightarrow Q$ must be true since it is an instance of a tautologous conditional statement form, as verified by each of the two methods presented in the previous section. In mathematics, a conditional statement that is determined to be true is called a *theorem*. Thus, we see that the conditional statement $[(P \rightarrow Q) \wedge (R \rightarrow S) \wedge (P \vee R) \wedge \overline{S}] \rightarrow Q$ is a theorem. It is clear that any valid argument is associated with a conditional statement that is a theorem. Also, if a conditional statement is a theorem, then this fact can be verified by showing that its associated argument is valid.

Section 3
Indirect Proof

Sometimes a conditional statement or argument is shown to be valid by use of an *indirect proof*. In an indirect proof, we assume that the conditional $p \rightarrow q$ is false. That is, we assume that $(\overline{p \rightarrow q})$ is true. Let r be a premise or other true statement, and recall that $(\overline{p \rightarrow q}) \Leftrightarrow (p \wedge \bar{q})$. Now, $p \wedge \bar{q}$ is true means that p is true and $\bar{q}$ is true. Hence, we can use both p and $\bar{q}$ as premises in a sequence of steps of a proof. If both r and $\bar{r}$ are justified as being in the sequence of steps, then $r \wedge \bar{r}$ would be true. But this is impossible, since $r \wedge \bar{r}$ is a contradiction. We conclude that the original assumption ($\overline{p \rightarrow q}$ is true) is false; that is, that $p \rightarrow q$ is true. The negation of the conclusion of a conditional statement must be used as a premise in a sequence of steps in an indirect proof. Furthermore, the conclusion of an indirect proof of this type is always a contradiction. The following examples should help clarify this idea.

Example 1: Consider the following argument. If Shafer took the job, then he must move. If he must move, then we will get new neighbors. Shafer took the job. Therefore, we will get new neighbors.

Let P, Q, and R stand for the appropriate statements. Then the conditional statement for this example is symbolized as $[(P \rightarrow Q) \wedge (Q \rightarrow R) \wedge P] \rightarrow R$. Assume the conditional is

Statements	Reasons
1. $\overline{R}$	1. Premise
2. $Q \to R$	2. Premise
3. $\overline{Q}$	3. 1, 2, and I.A.
4. $P \to Q$	4. Premise
5. $\overline{P}$	5. 3, 4, and I.A.
6. P	6. Premise
7. $P \land \overline{P}$	7. 5, 6, and Conjunction

Figure 3.6

false. That is, assume $\overline{[(P \to Q) \land (Q \to R) \land P] \to R}$ is true. The negation of the conditional is $[(P \to Q) \land (Q \to R) \land P] \land \overline{R}$. Hence we must show

$$\begin{array}{l} P \to Q \\ Q \to R \\ P \\ \overline{R} \\ \hline \therefore \text{ Contradiction.} \end{array}$$

The proof of this contradiction is given in Figure 3.6.

Since we have the contradiction of $P \land \overline{P}$ in Figure 3.6, we conclude that the conditional is true as stated; that is, $[(P \to Q) \land (Q \to R) \land P] \to R$ is a theorem, and the argument

$$\begin{array}{l} P \to Q \\ Q \to R \\ P \\ \hline \therefore R \end{array}$$

is valid. We point out that this theorem can be proved by a direct method also.

Example 2: Consider the indicated conditional statement $[(P \lor \overline{Q}) \land Q] \to P$.

The negation of this statement is $[(P \lor \overline{Q}) \land Q] \land \overline{P}$, and we

Statements	Reasons
1. $P \vee \overline{Q}$	1. Premise
2. $\overline{P}$	2. Premise
3. $\overline{Q}$	3. 1, 2, and A.A.
4. Q	4. Premise
5. $\overline{Q} \wedge Q$	5. 3. 4, and Conjunction

Figure 3.7

must show

$$\begin{array}{l} P \vee \overline{Q} \\ Q \\ \underline{\overline{P}\qquad\qquad} \\ \therefore \text{ Contradiction.} \end{array}$$

The indirect proof of this theorem is displayed in Figure 3.7.

The contradiction indicates that the conditional statement $[(P \vee \overline{Q}) \wedge Q] \rightarrow P$ is a theorem.

When you are trying to prove a theorem (prove an argument is valid), try the direct proof method first. If this method does not prove fruitful, then try the indirect proof method. If both methods fail, do not become disoriented. It could be that the argument is not valid or, equivalently, that the associated conditional statement is not a theorem.

Section 4
Invalid Arguments

An argument that is not valid is called an *invalid argument.* If you cannot show an argument valid, there can be one of two possible problems. First, you may not have had the correct insights to show the validity; second, the argument could be invalid. If, after repeated trials, you cannot show the validity of an argument, you may suspect that the argument is invalid. Of course, if the argument is invalid, there is no direct or indirect proof of its validity.

There are several methods for showing that an argument is invalid, only one of which will be given here. Since a valid argument must have an associated tautologous conditional statement form, it is reasonable to consider an argument invalid if this is not the case. That is, an argument can be shown to be invalid if its associated conditional statement form is not a tautology. Consider the following indicated argument:

$$\begin{array}{l} P \rightarrow Q \\ \underline{Q \qquad} \\ \therefore\ P \end{array}$$

The conditional statement form associated with this argument is $[(p \rightarrow q) \wedge q] \rightarrow p$. The truth table for this statement form is given in Figure 3.8. The truth table in Figure 3.8 is not a tautology; hence, the argument is invalid.

p	q	$p \rightarrow q$	$(p \rightarrow q) \wedge q$	$[(p \rightarrow q) \wedge q] \rightarrow p$
T	T	T	T	T
T	F	F	F	T
F	T	T	T	F
F	F	T	F	T

Figure 3.8

A conditional statement that is an instance of a non-tautologous conditional statement form is said to be an *invalid* statement.

Exercises 3.2, 3, 4

1. Prove the following statements which are theorems, or show the statement to be invalid.
a) $[(P \rightarrow Q) \wedge (Q \rightarrow R) \wedge (R \rightarrow S)] \Rightarrow (P \rightarrow S)$
b) $[(P \vee Q) \wedge \bar{Q} \wedge (P \rightarrow S)] \Rightarrow S$
c) $[(P \rightarrow Q) \wedge (Q \rightarrow R) \wedge P] \Rightarrow R$
d) $[(P \rightarrow Q) \wedge (Q \rightarrow R) \wedge (P \vee Q)] \Rightarrow (Q \vee R)$
e) $[(P \rightarrow Q) \wedge (Q \rightarrow R) \wedge \bar{Q} \wedge \bar{R}] \Rightarrow (\bar{P} \wedge \bar{Q})$
f) $[(\bar{Q} \vee \bar{S}) \wedge (P \rightarrow Q) \wedge (R \rightarrow S)] \Rightarrow (\bar{P} \vee \bar{R})$
g) $[(P \vee Q) \wedge (R \rightarrow \bar{P}) \wedge R] \Rightarrow Q$
h) $[\bar{Q} \wedge (P \rightarrow Q) \wedge (\bar{P} \rightarrow R)] \Rightarrow R$
i) $[P \wedge (P \rightarrow Q) \wedge R] \Rightarrow (P \wedge Q \wedge R)$
j) $\{[(P \wedge Q) \rightarrow R] \wedge (P \wedge Q)\} \Rightarrow R$
k) $[(P \rightarrow Q) \wedge \bar{P}] \Rightarrow \bar{Q}$

2. Provide a proof for the following valid arguments. Use the indicated abbreviations.

a) If metal is heated, then it will expand. The metal did not expand. Hence, the metal was not heated. (P: Metal is heated. Q: Metal will expand.)

b) If the door latch is broken, the house was broken into or the door latch was faulty. The door latch is broken and the door latch was not faulty. Therefore, the house was broken into. (P: The door latch is broken. Q: The house was broken into. R: The door latch was faulty.)

c) If the lawn is watered daily, it will be pretty and the water bill will be high. If the lawn is pretty, then the yard is enjoyed more. If the water bill is high, then our budget planning will be enjoyed less. The lawn is watered daily. Hence, the yard is enjoyed more and our budget planning is enjoyed less. (P: The lawn is watered daily. Q: The lawn will be pretty. R: The yard is enjoyed more. S: The water bill is high. T: Our budget planning will be enjoyed less.)

d) If Alpha toothpaste reduces cavities, then dentists have less work; and if fluoride is added to the water supply, adverse physical symptoms may develop. Either Alpha toothpaste reduces cavities or fluoride is added to the water supply. It is not the case that dentists have less work. Hence, adverse physical symptoms may develop. (P: Alpha toothpaste reduces cavities. Q: Dentists have less work. R: Fluoride is added to the water supply. S: Adverse physical symptoms may develop.)

e) Smith will go to work if he can get a job. He can meet his bills only if he gets union wages. Smith will not go to work, or it is not the case that he gets union wages. Therefore, he cannot get a job, or he cannot meet his bills. (P: Smith will go to work. Q: He can get a job. R: He can meet his bills. S: He gets union wages.)

f) Bill may use his educational opportunity to prepare himself for a professional career, or he may use his educational opportunity to prepare himself for a trade. Since Bill may use his educational opportunity to prepare himself for a professional career, or he may use his educational opportunity to prepare himself for a trade, or he may serve an apprenticeship in some area, then he will use his student years well and he will be a productive member of society, or he will waste his good preparation. He will not waste his good preparation. Therefore, he will be a productive member of society. (P: Bill may use his educational opportunity to prepare himself for a professional career. Q: He may use his educational opportunity to prepare himself for a trade. R: He may serve an apprenticeship in some area. S: He will use his student years well. T: He will be a productive member of society. U: He will waste his good preparation.)

Section 5

Use of Equivalent Forms and Derived Valid Argument Forms in Constructing Proofs

Suppose we try to prove that the following argument is valid.

I will not study mathematics unless I have a study schedule.

If I have a study schedule, then my leisure time will be reduced.

Therefore, if I study my mathematics, my leisure time will be reduced.

Using symbols, we express this argument as

$$\begin{array}{l} \overline{P} \vee Q \\ \underline{Q \rightarrow R} \\ \therefore\ P \rightarrow R \end{array}$$

Using only our valid argument forms, we are unable to supply a direct or indirect proof for the validity of this argument. Recall that $(\overline{p} \vee q) \Leftrightarrow (p \rightarrow q)$, as shown in the list of important equivalent forms in Section 14 of Chapter 2. Now, consider the sequence of statements in Figure 3.9.

We notice that Line 2 in Figure 3.9 is not a premise and does not follow from a valid argument form, as has been required by our definition of the proof of the validity of an argument. Recall that if two statements are equivalent, they have the same truth values. Hence, since $\overline{P} \vee Q$ is true and $\overline{P} \vee Q$ is equivalent to $P \rightarrow Q$, we know that $P \rightarrow Q$ is true, and it is clear that the sequence of steps in Figure 3.9 pro-

Statements	Reasons
1. $\overline{P} \vee Q$	1. Premise
2. $P \rightarrow Q$	2. 1 and instance of equivalent form (Conditional)
3. $Q \rightarrow R$	3. Premise
4. $P \rightarrow R$	4. 2, 3, and L.S.

Figure 3.9

vides a proof of the validity of the given argument. We see the plausibility of extending our definition of proof to allow for a line in a sequence of statements to be equivalent to one of the statements in a preceding line. We adopt this modification of our definition of a proof.

Consider the following indicated theorem.

$$[(\overline{P} \vee Q) \wedge (Q \rightarrow R) \wedge (R \rightarrow S) \wedge \overline{S}] \rightarrow \overline{P}$$

The sequence of statements in Figure 3.10 is suggested as a proof for this theorem. Notice that Line 3 is not a premise; it does not follow from previous lines by a valid argument form, and it is not an equivalent form for one of the preceding statements. Rather, it is an instance of the tautologous conditional statement form $[(\overline{p} \vee q) \wedge (q \rightarrow r)] \Rightarrow (p \rightarrow r)$ associated with the valid argument proved in Figure 3.9. This

Statements	Reasons
1. $\overline{P} \vee Q$	1. Premise
2. $Q \rightarrow R$	2. Premise
3. $P \rightarrow R$	3. 1, 2, and instance of a derived valid argument form
4. $R \rightarrow S$	4. Premise
5. $P \rightarrow S$	5. 3, 4 and L.S.
6. $\overline{S}$	6. Premise
7. $\overline{P}$	7. 5, 6, and I.A.

Figure 3.10

means that

$$\begin{array}{l} \bar{p} \vee q \\ \underline{q \rightarrow r} \\ \therefore\ p \rightarrow r \end{array}$$

is a valid argument form, although not one of the ten in Figure 3.1. We will call such valid argument forms *derived* valid argument forms. Note that we could have included the proof of

$$\begin{array}{l} \bar{P} \vee Q \\ \underline{Q \rightarrow R} \\ \therefore\ P \rightarrow R \end{array}$$

in the proof in Figure 3.10, but this would be needless effort. Again we extend our definition of a proof so that now we let a line of a proof be a premise or follow from previous lines by

1. A derived valid argument form.
2. A valid argument form in Figure 3.1.
3. An equivalent form in Section 14, Chapter 2.

Thus, the sequence of statements in Figure 3.10 consists of an acceptable proof for the validity of the indicated theorem.

The ten valid argument forms and ten equivalent forms mentioned previously are reasonably easy to use and are adequate for the construction of a proof for any valid argument in symbolic logic. This particular collection of valid argument forms and equivalent forms (or some variations of them) has become standard in logic and is the basis for the particular deductive structure with which we are working. This structure is an excellent model for conceptualizing the deductive method of reasoning.

The major concepts that we need for our deductive structure have now been presented. The next part of our work is the "icing on the cake." We have spent some time assembling our tools. Now that we have assembled an adequate number of them, we are ready for the fun of using them. To omit the material that follows would be like leaving the chocolate out of chocolate ice cream. To help the student summarize the material presented and to give order to his knowledge of this material, we present the information in Figure 3.11.

In Figure 3.11, we list the basic "rules" that indicate correct reasoning. In actually doing proofs, the student can consider himself in a "game" situation where the valid argument forms and equivalent forms are the basic "rules." As the student proves arguments valid, the associated derived valid argument forms can be thrown back into the system to be used as "rules." Constructing the proof for a theorem or valid argument is much like putting a puzzle together. One must arrange the sequence of statements in such a way as to use *only* the "rules" as justification for a particular sequence. For any proof in this chapter, only the valid argument forms and equivalent forms will need to be used. However, if a previously derived valid argument form fits the situation, the proof may be shortened by using that derived valid argument form.

Consider the following indicated theorem.

$$\{(P \wedge Q) \wedge [P \rightarrow (Q \rightarrow R)] \wedge (R \rightarrow T) \wedge (S \vee \overline{T})\} \rightarrow S$$

The first step in preparing to prove this theorem is to write it as an argument:

$$\begin{array}{l} P \wedge Q \\ P \rightarrow (Q \rightarrow R) \\ R \rightarrow T \\ S \vee \overline{T} \\ \hline \therefore\ S \end{array}$$

The second step is to draw a diagram (at least until you are at ease with proof problems) with a statements column and a reasons column, as illustrated in Figure 3.12. Now, we are down to the proof itself. How do you start? You will be happy to know there is no fixed procedure for doing a proof. You may exercise your own creativity. The important thing is to write something; that is, *begin*! Many students are hesitant about starting a deductive procedure unless they see the whole pattern right from the start. However, a proof is usually constructed just one step at a time, with each step

VALID ARGUMENT FORMS	EQUIVALENT FORMS
Law of Detachment $p \rightarrow q$ p $\therefore q$	DeMorgan's Laws $(\overline{p \wedge q}) \leftrightarrow (\bar{p} \vee \bar{q})$ $(\overline{p \vee q}) \leftrightarrow (\bar{p} \wedge \bar{q})$
Indirect argument $p \rightarrow q$ $\bar{q}$ $\therefore \bar{p}$	Commutation $(p \vee q) \leftrightarrow (q \vee p)$ $(p \wedge q) \Leftrightarrow (q \wedge p)$
Law of syllogism $p \rightarrow q$ $q \rightarrow r$ $\therefore p \rightarrow r$	Conditional $(p \rightarrow q) \Leftrightarrow (\bar{p} \vee q)$ $(p \vee q) \Leftrightarrow (\bar{p} \rightarrow q)$
Alternative argument $p \vee q$, $\bar{q}$, $\therefore p$ or $p \vee q$, $\bar{p}$, $\therefore q$	Double negation $p \Leftrightarrow \bar{\bar{p}}$
Simplification $p \wedge q$, $\therefore p$ or $p \wedge q$, $\therefore q$	Exportation $[(p \wedge q) \rightarrow r] \Leftrightarrow [p \rightarrow (q \rightarrow r)]$
Absorption $p \rightarrow q$ $\therefore p \rightarrow (p \wedge q)$	Distributivity $[p \wedge (q \vee r)] \Leftrightarrow [(p \wedge q) \vee (p \wedge r)]$ $[p \vee (q \wedge r)] \Leftrightarrow [(p \vee q) \wedge (p \vee r)]$

Figure 3.11

Constructive dilemma

$p \rightarrow q$
$r \rightarrow s$
$p \vee r$
$\therefore q \vee s$

Destructive dilemma

$p \rightarrow q$
$r \rightarrow s$
$\overline{q} \vee \overline{s}$
$\therefore \overline{p} \vee \overline{r}$

Addition

p
$\therefore p \vee q$

Conjunction

p
q
$\therefore p \wedge q$

Equivalence

$(p \leftrightarrow q) \Leftrightarrow [(p \rightarrow q) \wedge (q \rightarrow p)]$
$(p \leftrightarrow q) \Leftrightarrow [(p \wedge q) \vee (\overline{p} \wedge \overline{q})]$

Association

$p \vee (q \vee r) \Leftrightarrow (p \vee q) \vee r$
$p \wedge (q \wedge r) \Leftrightarrow (p \wedge q) \wedge r$

Contrapositive

$(p \rightarrow q) \Leftrightarrow (\overline{q} \rightarrow \overline{p})$

Reduction

$p \Leftrightarrow (p \vee p)$
$p \Leftrightarrow (p \wedge p)$

DERIVED VALID ARGUMENT FORMS

Argument:

———
——— Premises
———

∴ ——— Conclusion

Proof:

Statements	Reasons
1. ———	1. ———
2. ———	2. ———
.	.
.	.
.	.

Deduction in Symbolic Logic

Figure 3.11 continued

Statements	Reasons
1.	1.
2.	2.
.	.
.	.
.	.

Figure 3.12

suggesting the next. For a proof of the theorem indicated just above, see Figure 3.13.

Familiarity with the "rules" makes it easier for you to see the correct step. You scan your mind to pick out the appropriate valid argument form or equivalent form that will justify a particular line. Or, perhaps it's the other way around. You pick the "rule" that suggests the step. In any case, the more you practice, the more agile you will become in constructing proofs.

Suppose you live in City A and decide to visit Place B. If you have not traveled to Place B before, you must use a map to select a route that will take you to Place B. As you are selecting a route, you will possibly consider some that will not arrive at Place B. Some will be inconvenient; some will be scenic; some will be better fitted to a particular type

Statements	Reasons
1. $P \rightarrow (Q \rightarrow R)$	1. Premise
2. $(P \wedge Q) \rightarrow R$	2. 1 and Exp.
3. $P \wedge Q$	3. Premise
4. R	4. 2,3, and L.D.
5. $R \rightarrow T$	5. Premise
6. T	6. 4,5, and L.D.
7. $S \vee \bar{T}$	7. Premise
8. $\bar{\bar{T}}$	8. 6 and D.N.
9. S	9. 7,8, and A.A.

Figure 3.13

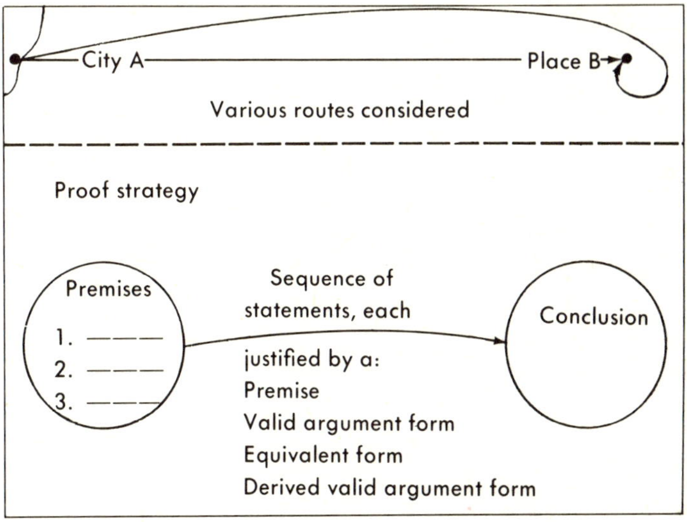

Figure 3.14

of transportation than others; some will be direct; etc. After some thought, you make a decision as to what means of travel and what route best fits your needs. In your planning, the paramount concern is to arrive at Place B. Secondly, you concern yourself with associated matters, such as the weather, the time involved, etc. The analogy to deductive reasoning is illustrated in Figure 3.14. Your main goal is to select a proof pattern that does the job. In the process you may have to reassess your progress; and, after a proof is found, you may want to find a clearer, shorter proof.

Deductive reasoning in any structure is a similar process, although the particular substance of the structures may vary. Once a proof is discovered, anyone who can follow instructions can determine whether or not the proof is correct. All he must do is check to see if each line in the proof is a premise or is justified by a valid argument form, equivalent form, or derived valid argument form. Although checking a proof can be reduced to a mechanical procedure, finding a proof is a different matter. Mathematicians have shown that for many important mathematical theories, there is no mechanical method for finding a proof. Finding a proof is a creative act. As is the case with most creative activities, it usually requires much practice, insight, tenacity, and something called "luck." For most mathematicians, the beauty of participating in the creative act of proof-finding never fades nor does the appreciation of the beauty of a proof someone else discovers. In fact, the more you participate in the activity, the more rewarding it is.

The following exercises will give you an opportunity to become familiar with the deductive method and to exercise your creativity in finding appropriate proof patterns.

Exercises 3.5

Provide a proof for the following indicated theorems.

1. $[(P \vee \overline{Q}) \wedge Q] \Rightarrow P$
2. $[(\overline{P} \rightarrow Q) \wedge \overline{Q}] \Rightarrow P$
3. $[(P \rightarrow Q) \wedge (Q \rightarrow R)] \Rightarrow (\overline{R} \rightarrow \overline{P})$
4. $[(P \rightarrow Q) \wedge (P \rightarrow R)] \Rightarrow [P \rightarrow (Q \wedge R)]$
5. $[P \rightarrow (Q \wedge R)] \Rightarrow [(P \rightarrow Q) \wedge (P \rightarrow R)]$
6. $[(P \vee Q) \wedge (P \rightarrow R) \wedge (Q \rightarrow R)] \Rightarrow R$
7. $[(P \wedge \overline{Q}) \rightarrow \overline{P}] \Rightarrow (P \rightarrow Q)$
8. $[(P \wedge \overline{Q}) \rightarrow (R \wedge \overline{R})] \Rightarrow (P \rightarrow Q)$
9. $\overline{P} \Rightarrow (\overline{P \wedge Q})$
10. $\{[\overline{P} \rightarrow (Q \rightarrow \overline{R})] \wedge (\overline{P} \vee S) \wedge (\overline{T} \rightarrow Q) \wedge \overline{S}\} \Rightarrow (R \rightarrow T)$
11. $[(P \rightarrow Q) \wedge (R \rightarrow Q)] \Rightarrow [(P \vee R) \rightarrow Q]$
12. $\{(P \rightarrow Q) \wedge (\overline{Q \vee R}) \wedge [(\overline{P} \wedge \overline{R}) \rightarrow S]\} \Rightarrow S$
13. $\{[P \rightarrow (Q \rightarrow R)] \wedge (\overline{S} \vee P) \wedge Q\} \Rightarrow (S \rightarrow R)$
14. $[(P \leftrightarrow Q) \wedge (R \rightarrow S) \wedge (P \vee R) \wedge (P \rightarrow \overline{S})] \Rightarrow (Q \leftrightarrow \overline{S})$
15. $\{[(P \wedge Q) \rightarrow S] \wedge [(Q \rightarrow S) \rightarrow (R \vee T)] \wedge (\overline{R} \wedge \overline{T})\} \Rightarrow \overline{P}$

Section 6
Symbolization of Theorems

In Exercises 3.2, 3, 4, Problem 2, the symbols P, Q, R, etc., were specifically designated, which aided in symbolizing the arguments. Now, in the next set of exercises you will be asked to make the appropriate designation of symbols.

One of the keys to determining the validity of a particular argument is to translate the argument accurately from word symbols into logic symbols. Consider the following argument:

If you need a powerful rifle or are target shooting at long distances, then you are preparing to hunt large game. If you do not need a powerful rifle, you are not preparing to hunt large game. Since you do not need a powerful rifle, you are not target shooting at long distances.

In translating an argument into symbols, we suggest that you underline the simple statements in the argument and put the appropriate statement symbols above these simple statements. Thus, you would have:

P Q
If you need a powerful rifle or you are target shooting at long

R
distances, then you are preparing to hunt large game. If you do

Statements	Reasons
1. $(P \vee Q) \rightarrow R$	1. Premise
2. $\overline{R} \rightarrow \overline{(P \vee Q)}$	2. 1 and equivalent form Contra.
3. $\overline{P} \rightarrow \overline{R}$	3. Premise
4. $\overline{P} \rightarrow \overline{(P \vee Q)}$	4. 2, 3, and L.S.
5. $\overline{\overline{P}} \vee \overline{(P \vee Q)}$	5. 4 and equivalent form Cond.
6. $P \vee \overline{(P \vee Q)}$	6. 5 and equivalent form D.N.
7. $P \vee (\overline{P} \wedge \overline{Q})$	7. 6 and equivalent form De.M.
8. $(P \vee \overline{P}) \wedge (P \vee \overline{Q})$	8. 7 and equivalent form Dist.
9. $P \vee \overline{Q}$	9. 8 and Simp.
10. $\overline{P} \rightarrow \overline{Q}$	10. 9 and equivalent form Cond.

Figure 3.15

$\overline{P}$ $\overline{R}$
not need a powerful rifle, you are not preparing to hunt large
$\overline{P}$
game. Since you do not need a powerful rifle, you are not target
$\overline{Q}$
shooting at long distances.

You can symbolize the above argument in theorem form as $\{[(P \vee Q) \rightarrow R] \wedge (\overline{P} \rightarrow \overline{R})\} \rightarrow (\overline{P} \rightarrow \overline{Q})$ and in argument form as

$$\begin{array}{l} (P \vee Q) \rightarrow R \\ \overline{P} \rightarrow \overline{R} \\ \hline \therefore\ \overline{P} \rightarrow \overline{Q} \end{array}$$

A proof is given in Figure 3.15.

Exercises 3.6

To help you symbolize the following arguments, we have indicated letters that are associated with key words in the component statements. These letters should be substituted for their associated statements. Provide a proof that establishes the validity of the following arguments.

1. The stereo does not play. Hence, it is not the case that the stereo plays and we can use it at the party. (S, P)

2. It is false that Smith bought either a Corvette or a Mercedes. So, Smith did not buy a Corvette. (C, M)

3. If I register for classes today, then I will not have to pay a late registration fee. I can register for classes today if I have met all preregistration requirements. Not having sent my transcript in on time is a sufficient condition for having to pay a late registration fee. Hence, I have met all preregistration requirements only if I have sent my transcript in on time. (R, P, M, T)

4. Our team can expect to win the ball game only if the team will practice long hours. If our team members will have good discipline, then the team will practice long hours. Our team can expect to win the ball game or our team members will have good discipline, provided the coach is a good motivator. The coach is a good motivator. Therefore, the team will practice long hours. (W, P, D, C)

5. I want to lose weight or my clothes will not fit well. I cannot lose weight whenever I am not active or I cannot afford a dietician's advice. Hence, my clothes will fit only if I can be active and if I can afford a dietician's advice. (L, C, A, D)

6. Jane is happy only if Zilch takes her to parties whenever he goes to parties. The romance is enchanting if Zilch takes her to parties or he does not go to parties. The romance is not enchanting. Hence, Jane is unhappy. (J, Z, P, E)

7. If I buy a new Ford or a new Chevrolet, then I will get a good deal and own a fine car. If I own a fine car or I rent a new car, then I will take a vacation. So, I will take a vacation if I buy a new Ford. (F, C, D, O, R, V)

8. If flying is fun, then I will visit Chicago or Phoenix or St. Louis. If flying is expensive and I visit Chicago, then I will stay a short time. Flying is fun and flying is expensive. I do not have time to visit either Phoenix or St. Louis. Hence, I will stay a short time. (F, E, C, P, S, T)

9. If Mrs. Fleetfoot knows of the incident, then Mrs. Fleetfoot told Mrs. Wantstoknow, and Mr. Fleetfoot talked to the maid only if Mr. Fleetfoot called home today. If Mrs. Fleetfoot told Mrs. Wantstoknow, then the incident is common knowledge, and Mr. Fleetfoot was told of the incident by the maid if he called home today. It is false that the incident is common knowledge and that Mr. Fleetfoot was told of the incident by the maid. Mr. Fleetfoot talked to the maid if Mrs. Fleetfoot knows of the incident. Therefore, Mrs. Fleetfoot does not know of the incident. (K, W, T, H, C, M)

10. If Jane goes to the market, then she buys pork chops or she buys beefsteak, provided she has plenty of money. Jane fixes fried pork chops whenever she buys pork chops, and Jane fixes barbecued beefsteak whenever she buys beefsteak. Jane has gone to the market and she has plenty of money. Jane did not fix barbecued beefsteaks. Therefore, Jane fixed fried pork chops. (J, C, S, M, F, B)

Section 7
Mathematics

We have been studying many interesting topics in mathematics. Up to this point we have not asked the question "What is mathematics?" A good reason for not asking it is because we do not know the answer! In fact, this is one of the intriguing aspects of studying mathematics. We, as most of you, are giving intense consideration to a subject area that we cannot encapsulize in a definition. There are many ways partially to describe mathematical endeavors. Mathematics is the study of abstract systems. It is also the language of science. Mathematics is a process as well as a product. Mathematics is creative and deductive. It is a symbolic language; it is a cultural product; it is a study of numbers; it is a study of algorithms; etc. Mathematics is all of these things, and yet all of these words and phrases cannot provide an acceptable definition of mathematics. Do we give up in despair because a definition of mathematics eludes us? No. We will admit that mathematics is at once awesomely majestic and yet enticingly simple. We will do what one does in so many other endeavors—pretend that we do know what mathematics is and proceed to study it. After involving ourselves in its study for awhile, we will each form our own opinion about what mathematics is. Yet, none of us would be so audacious as to suppose that our understanding is complete enough to attempt to define mathematics.

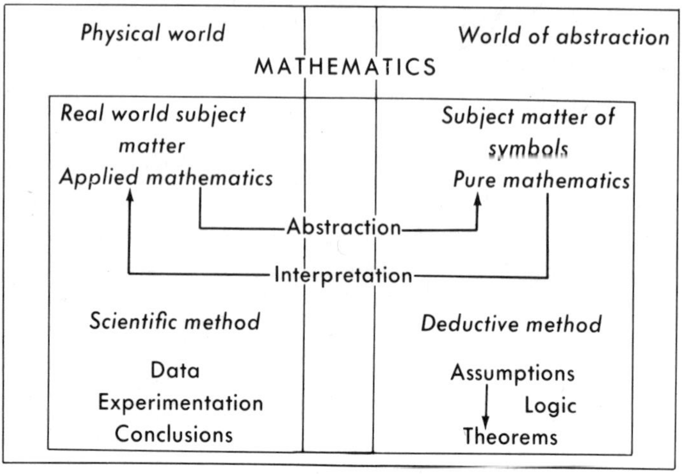

Figure 3.16

Mathematics that is purely abstract is generally referred to as "pure mathematics," while mathematics that deals primarily with physical reality is called "applied mathematics." Applied mathematics and pure mathematics form distinct branches of mathematical endeavor. The study of physics is an example of applied mathematics, while the study of symbolic logic is an example of pure mathematics. Pure and applied mathematics are interrelated to the extent that each prods and extends the other. Applied mathematics essentially deals with real world relationships, while pure mathematics essentially deals with relationships of symbols in an abstract system. The diagram in Figure 3.16 illustrates in a rudimentary way the interrelatedness of applied and pure mathematics. The rectangles for the physical world and the world of abstraction overlap because it is sometimes difficult to separate the two. Instead of thinking of applied and pure mathematics as discrete and separable, think of a continuum of interrelationship between the two.

Section 8
Deductive Systems

The preceding section was included to help the student understand the encompassing nature of mathematics and the interrelatedness of physical world mathematics with abstract. Our emphasis in this text is upon the abstract. This emphasis provides us with an opportunity to become aware of many important generalizations in mathematics, and it gives us practice in using the deductive method. A *deductive system* is composed of undefined terms, defined terms, assumptions, derived statements, and the principles of logic. In Chapter 1 we discussed undefined terms and definitions. The words *axiom* and *postulate* are used interchangeably with *assumption.* A derived statement or provable statement in a deductive system is called a theorem.

Our work up to this point provides a foundation for work in any deductive system from two points of view. First, the method of deduction used so far is analogous to the method that is used in any deductive system. Second, the tautologies of symbolic logic are the statement forms that determine correct deductive procedures in any system. A *proof* of a theorem in a deductive system consists of a sequence of statements in which each statement is a premise, or is justifiable by an axiom, a definition, a previously proved theorem, or a principle of logic.

The concepts of logic and set theory have been presented intuitively. Although they can be presented in a deductive system, it is outside the purpose of this text to do so. The concepts of logic and set theory developed thus far will be used in constructing deductive systems and in providing models or interpretations for those deductive systems in the real world.

A few remarks need to be made regarding desirable properties for a group of axioms. First, axioms should be *minimal* in nature. That is, if we have a list of axioms in a deductive system and one of them can be deduced from the others, then that one axiom is unnecessary. It should be taken out of the list of axioms and be derived from the other axioms. Second, we do not want the embarrassing situation of being able to prove a statement both true and false in our deductive system. Hence, we desire a *consistent* set of axioms; that is, a set that will not allow for statements P and P both to be provable. Third, a system of axioms should possess *completeness*. We need to be able to show that any statement in a deductive system is a theorem or that the addition of this statement to the set of axioms would render the set of axioms inconsistent.

Section 9
Concluding Remarks

Now that we have the concepts of sets, logic, and deductive reasoning, we are ready to investigate some basic deductive systems or mathematical structures. These systems are cen-

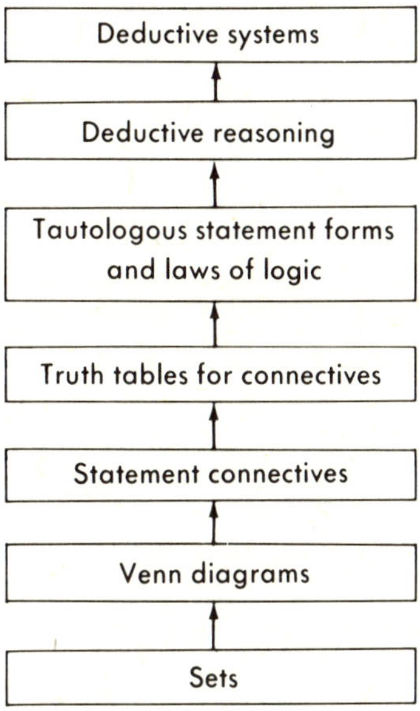

Figure 3.17

tral to the study of mathematics, and hopefully will be intellectually satisfying, interesting, and informative.

The principles of deductive reasoning set forth in this chapter are, in the authors' opinion, some of the most beautiful creations that the mind of man has ever produced. We have only presented standard information in symbolic logic and deductive reasoning. There are many more interesting facets in the study of logic. In the formal study of logic, man's confidence in the usefulness of deductive reasoning is extended, and yet some of the results of the last fifty years have made us aware of how limited the deductive process is. We encourage you to read further in the area of logic if you like challenging and enchanting reading.

We provide the diagram in Figure 3.17 to help highlight our development to this point and to help you relate the various components of our study to each other.

References Chapter Three

Delong, Howard, *A Profile of Mathematical Logic* (Reading, Pa.: Addison-Wesley Publishing Company, 1970), pp. 191–229.

Fremont, Herbert, *How to Teach Mathematics in Secondary Schools* (Philadelphia: W. B. Saunders Company, 1969), pp. 3–12.

Laatsch, Richard, *Basic Algebraic Systems: An Introduction to Abstract Algebra* (San Francisco: McGraw-Hill Book Company, 1968), pp. 195–201.

Spitznagel, Edward L., Jr., *Selected Topics in Mathematics* (Dallas: Holt, Rinehart and Winston, Inc., 1971), pp. 1–5.

Stein, Sherman K., *Mathematics, The Man-Made Universe* (San Francisco: W. H. Freeman and Company, 1963), pp. 272–277.

Newman, James R., ed., *The World of Mathematics* (New York: Simon and Schuster, 1956), vol. III, pp. 1647–1667; vol. IV, pp. 2099–2123.

CHAPTER FOUR

Relations and Functions

The material in this chapter is basic to most mathematical endeavors. The concepts presented will provide us with some powerful notation and terminology that will facilitate our study of deductive mathematical systems.

We are somewhat in the position of one at the base of a mountain who, seeing what appears to be the summit, sets out to climb to the top. But upon reaching what appeared to be the top, he realizes that the actual summit is farther up and could not be seen from the base of the mountain. So it is with our development. By ascending a short distance further, you will reach a vantage point from which it is possible to perceive a more comprehensive view of mathematics.

In Chapter 1, we introduced the notions of ordered pair and Cartesian product. In our use of these concepts, we provided a set-theoretic discussion of multiplication of cardinal numbers. The concepts of ordered pair and Cartesian product will provide a means for discussing relations and functions in this chapter.

Section 1
Relations

How would you define the word *relation?* Write your definition. Now see if you can state it in set language. As you read the following material, you will probably see that the intuitive idea of *relation,* which you have expressed, will be closely associated with the precise statements we will make regarding the mathematical meaning of the term.

Consider the set of Presidents A = {Washington (W), Lincoln (L), Nixon (N)}. With regard to the sequence of the terms of these presidents, when elements from set A are substituted for x and y in the expression "x preceded y," construct an ordered pair of the form (x, y). In this manner, the expression "x preceded y" will generate the set of ordered pairs B = {(W, L), (W, N), (L, N)}. The expression, "x preceded y" is an example of a sentence that determines a relation on set A. Notice that the set of ordered pairs {(W, L), (W, N), (L, N)} is a subset of the Cartesian product A × A = {(W, W), (W, L), (W, N), (L, W), (L, L), (L, N), (N, W), (N, L), (N, N)}. The word "preceded" tells us how the coordinates of the ordered pairs of set B are related. Notice that the set B of ordered pairs completely describes the rule of correspondence specified by the word "preceded," so that in each ordered pair the first coordinate is associated with its corresponding second coordinate according to this rule.

The set C = {(L, N), (L, W), (N, W)} is an example of a

different relation that exists on set A. With regard to the order in the alphabet of the first letter of the names in set A, the expression "x precedes y" gives rise to set C.

From the set A × A, suppose we select the subset of ordered pairs D = {(L,W), (N,W), (N,L)}. Can you think of a rule of correspondence that is determined by the set D? There are no doubt several possibilities, but the one the authors have in mind is "the name with fewer letters precedes the name with more letters." The ordered pairs in set D indicate a correspondence between the elements of set A, and the expression "the name with fewer letters precedes the name with more letters" is a rule that describes the correspondence between the elements of set A as indicated in set D.

On the basis of the above discussion, we see that a subset of the Cartesian product A × A is determined by a "sentence" or "rule," that describes how elements of the set A are related or correspond. Also, the ordered pairs of a subset of A × A determine a "rule" by which the elements of the set A correspond.

So far we have only considered relations on a single set. That is, both coordinates of the ordered pairs have been from the same set. A relation may exist between two different sets, however. That is, the first coordinate of each ordered pair comes from one of the sets and the second coordinate comes from the other. For example, let A = {car, plane, boat} and let B = {drive, fly, sail}. The words in set A are related to the words in set B in an obvious manner. This relationship can be completely expressed by the rule of correspondence indicated by the ordered pairs M = {(car, drive), (plane,fly), (boat,sail)}. Notice that M is a subset of the Cartesian product A × B. The ideas presented in the preceding discussion are incorporated into the following definitions.

Definition: A set is a *relation* if and only if it is a set of ordered pairs.

Definition: A *relation from set A to set B* is a subset of A × B.

Definition: For a set A, a subset of $A \times A$ is a *relation on set A.*

Example 1: Let $A = \{1, 2, 3\}$ and $B = \{a, b, c\}$. Then, $C = \{(1,a), (2,b), (3,c)\}$ is a relation from set A to set B.

Example 2: Let $Z = \{\ldots,-3,-2,-1,0,1,2,3,\ldots\}$. We see that the rule of correspondence "square of" is clearly shown by the subset C of $Z \times Z$, $C = \{\ldots,(-4,16), (-3,9), (-2,4), (-1,1), (0,0), (1,1), (2,4), (3,9), (4,16),\ldots\}$.

Example 3: Consider the students in the set A = {Sam (S), Fred (F), Judy (J)}, and the courses in the set B = {English (E), Golf (G), Physics (P), History (H), Mathematics (M), Biology (B)}.

The subset of ordered pairs of $A \times B$ that expresses the relation between the students in set A and the courses in set B in which they are enrolled is as follows:

$$C = \{(S,E), (S,G), (S,P), (S,H), (F,E), (F,P), (F,M), (J,E), (J,H), (J,M), (J,B)\}.$$

From set C, we can determine that Sam is studying English, Golf, History, and Physics; Fred is studying English, Physics, and Math; and Judy is studying English, History, Biology, and Math.

In general, a relation is designated by the letter R, and we indicate that a particular ordered pair (a,b) belongs to R by writing $(a,b) \in R$. For any relation R, there are two sets that are of particular interest. These sets are defined in the following manner:

Definition: Let A and B be sets, and let R be a subset of $A \times B$. The *domain* of the relation R is the set of all first coordinates of the ordered pairs in R. We denote this set $\mathfrak{D}(R)$, and write symbolically

$$\mathfrak{D}(R) = \{x \mid (x,y) \in R\}.$$

Definition: Let A and B be sets, and let R be a subset of $A \times B$. The *range* of the relation R is the set of all second coordinates of the ordered pairs in R. We denote this set $\mathcal{R}(R)$, and write symbolically

$$\mathcal{R}(R) = \{y \mid (x,y) \in R\}.$$

Note that $\mathcal{D}(R) \subset A$ and $\mathcal{R}(R) \subset B$.

Example 4: Let R = {(a,red), (b,white), (c,blue), (d,green)}. Then $\mathcal{D}(R)$ = {a,b,c,d} and $\mathcal{R}(R)$ = {red, white, blue, green}.

Example 5: Let A = {water, air, land, sand, clouds}. Let B = {boat, plane, car, swim, run}. If R = {(water,boat), (air,plane), (land,car), (water,swim)}, then $\mathcal{D}(R)$ = {water, air, land} and $\mathcal{R}(R)$ = {boat, plane, car, swim}. We see that $\mathcal{D}(R) \subset A$ and $\mathcal{R}(R) \subset B$.

Consider the relation $R = \{(a,b) \mid a = b + 3$ where a and b are integers$\}$. Notice that the way a and b are related is given in terms of a formula. In many cases a formula is a convenient and economical way to indicate a rule of correspondence for the ordered pairs of a relation.

A special type of relation that is of prime importance in mathematics is an equivalence relation.

Definition: A relation R defined on a set A is said to be an *equivalence relation* if and only if

1) R is *reflexive;* i.e., for every $a \in A$, $(a,a) \in R$.

2) R is *symmetric;* i.e., if $(a,b) \in R$, then $(b,a) \in R$.

3) R is *transitive;* i.e., if $(a,b) \in R$ and $(b,c) \in R$, then $(a,c) \in R$.

Example 6: Let A = {1,2,3}. The relation R = {(1,1), (2,2), (3,3), (1,2), (2,1), (2,3), (3,2), (1,3), (3,1)} is an equivalence relation.

Example 7: If Z is the set of integers, then $R = \{(x,y) \mid x - y = 4n$ where $x, y, n \in Z\}$ is an equivalence relation since:

1) Letting $n = 0$, we see that $x - x = 4 \cdot 0$ and, hence, $(x,x) \in R$ for all $x \in Z$.

2) If $(x,y) \in R$, then $x - y = 4n$ where n is an integer. But then, $y - x = -4n = 4(-n)$. So since n is an integer, this means that $-n$ is also an integer and, hence, $(y,x) \in R$.

3) If $(x,y) \in R$ and $(y,z) \in R$, then $x - y = 4n$ and $y -$

$z = 4m$ where n and m are both integers. By adding, we have

$$(x - y) + (y - z) = 4n + 4m$$
$$x - z = 4n + 4m$$
$$x - z = 4(n + m)$$

Thus, $(x,z) \in R$, since $n + m$ will always be an integer when n and m are both integers.

Exercises 4.1

1. Give a rule of correspondence for the following relations.
a) R = $\{(1,2), (2,3), (3,4), (4,5)\}$
b) R = $\{(5,\frac{5}{2}), (6,3), (7,\frac{7}{2}), (8,4)\}$
c) R = {(Monday, Mo), (Tuesday, Tu), (Wednesday, We), (Thursday, Th), (Friday, Fr), (Saturday, Sa), (Sunday, Su)}
d) R = $\{(1,6), (2,8), (3,10), (4,12), (5,14)\}$
e) R = {(apples, red), (bananas, yellow), (plums, purple)}
f) R = {(bacon, pig), (rump roast, cow), (mutton, sheep)}
g) R = {(cold, hot), (night, day), (asleep, awake), (alive, dead)}

2. Let A = {Maine, Texas, Wyoming, Indiana, Oregon}. Give the relation $R = \{(x,y) \mid (x,y) \in A \times A\}$ that the given rule of correspondence determines.
a) "x is farther west than y."
b) "The name x has fewer letters than the name y."
c) "Alphabetically, the name x comes before the name y."
d) "The name x has the same number of letters as the name y."

3. Let $A = \{1,2,3,4,5,6,7,8,9,10\}$. Give the relation $R = \{(x,y) \mid (x,y) \in A \times A\}$ that the given rule of correspondence determines.
a) $y = x + 4$
b) $3x = y$
c) $y = (\frac{3}{2})x + 2$
d) "y is one less than twice x."
e) "y is one half of the sum of x and 3."

4. Which of the following are relations?

a) $(2,4)$
b) $\{(2,4)\}$
c) $\{\{(2,4)\}\}$
d) $\{(2,4), (1,5)\}$
e) $\{\{2,4\}, \{1,5\}\}$
f) $\{(2,\{4\}), (1,\{5\})\}$
g) $\{(-3,6), \{-5,7\}\}$
h) $\{\{\{(-3,6), (-5,7)\}\}\}$

5. For each of the following relations, state the domain of the relation and the range of the relation.
a) {(red, apple), (yellow, banana), (green, lime)}
b) $\{(7,3), (-2,12), (57,-4), (83,97)\}$
c) $\{(a,b) \mid a \in \{0,1,2,3\} \text{ and } b = 3a + 8\}$

6. In each of the following, a relation R is defined on a given set. Which of these relations are equivalence relations?

a) Let A be any set. $R = \{(P,Q) \mid P \subset A \text{ and } Q \subset A \text{ and } P \equiv Q\}$.
b) Let A be the set of all cars. Let $(x,y) \in R$ if and only if x is the same make as y.
c) Let A be the set of all people. Let $(x,y) \in R$ if and only if x was born in the same year as y.
d) Let A be the set of all people. Let $(x,y) \in R$ if and only if x is the twin of y.
e) Let $A = \{4,5,6,7\}$. $R = \{(4,7), (6,6), (5,4), (7,7), (5,7), (5,5), (7,4), (4,5), (4,4)\}$.
f) Let A = {house, car, TV}. R = {(house, house), (car, car), (TV, TV)}.
g) Let Z = the set of integers. $R = \{(x,y) \mid x + y = 3n$ where $x, y, n \in Z\}$.
h) Let Z = the set of integers. $R = \{(x,y) \mid x - y = \frac{n}{3}$ where $x, y, n \in Z\}$.
i) Let Z = the set of integers. $R = \{(x,y) \mid x - y$ is less than 0 where $x, y \in Z\}$.
j) Let Z = the set of integers. $R = \{(x,y) \mid x - y = xn$ where $x, y, n \in Z\}$.
k) Let Z = the set of integers. $R = \{(x,y) \mid x - y = 3n$ where $x, y, n \in Z\}$.
l) Let Z = the set of integers. $R = \{(x,y) \mid \frac{x}{y} = 3n$ where $x, y \in Z$ and $n \in Q\}$.

7. List all the relations on $\{x,y\}$. Then list those relations that are:
a) reflexive
b) transitive
c) symmetric
d) equivalence relations

Section 2
Inverse Relations

Let R = {(triangle, 3), (quadrangle, 4), (pentagon, 5)}. Then, $\mathfrak{D}$(R) = {triangle, quadrangle, pentagon} and $\mathfrak{R}$(R) = {3, 4, 5}. A sentence that expresses a rule of correspondence for R is "x is a geometric figure that has y vertices." Consider the sentence "A geometric figure with w sides is a z" where $w \in \mathfrak{R}(R)$ and $z \in \mathfrak{D}(R)$. This sentence determines the set of ordered pairs A = {(3, triangle), (4, quadrangle), (5, pentagon)}. Notice that A is a set of ordered pairs and, hence, a relation. It is the relation that we get from R by interchanging the first and second coordinates of every ordered pair in R.

Given a relation from set A to set B, it is desirable in many mathematical and natural settings to consider the corresponding relation from set B to set A. We express this notion precisely in the following definition.

Definition: Let R be a relation. The relation {(b,a) | (a,b) $\in$ R} is the *inverse relation* of R, denoted R^{-1}. That is, R^{-1} = {(b,a) | (a,b) $\in$ R}

Example 8: Let A = {Bill, Chuck, Frank, Jack}, B = {golf, baseball, football, wrestling, tennis} and R = {(Bill, golf), (Frank, baseball), (Frank, wrestling), (Jack, tennis)}. Note that $\mathfrak{D}$(R) = {Bill, Frank, Jack} and $\mathfrak{D}(R) \subset A$. Also note that $\mathfrak{R}$(R) = {golf, baseball, wrestling, tennis} and $\mathfrak{R}(R) \subset B$. Now, R^{-1} = {(golf, Bill), (baseball, Frank), (wrestling, Frank), (tennis, Jack)}. Then,

$\mathfrak{D}(R^{-1})$ = {golf, baseball, wrestling, tennis} and $\mathfrak{R}(R^{-1})$ = {Bill, Frank, Jack}. We see that $\mathfrak{R}(R) = \mathfrak{D}(R^{-1})$ and $\mathfrak{D}(R) = \mathfrak{R}(R^{-1})$, which we would expect upon reflection of the definition of R^{-1}.

Example 9: Suppose the rule of correspondence for the relation R from set A to set B is given by the formula $b = 2a + 3$ and $\mathfrak{D}(R) = \{1,3,5\}$. Then, $R = \{(1,5), (3,9), (5,13)\}$. Also, $R^{-1} = \{(5,1), (9,3), (13,5)\}$. Can you determine the rule of correspondence for the components of the ordered pairs in R^{-1}? The rule of correspondence is $a = \frac{b - 3}{2}$. Notice that $\mathfrak{D}(R^{-1}) = \mathfrak{R}(R) = \{5, 9, 13\}$. That is, we let $b = 5$ and then $a = \frac{5 - 3}{2} = 1$. Similarly, when $b = 9$, then $a = 3$ and for $b = 13$, $a = 5$. In this manner we determine the ordered pairs in R^{-1}.

The exercises will provide you with experience in determining a rule of correspondence associated with R^{-1} when given a rule of correspondence associated with R.

Exercises 4.2

1. For the relation $R = \{(x,1), (y,2), (z,3)\}$, what is R^{-1}?

2. Construct an equivalence relation on the set $A = \{1\}$. Is the inverse of this relation an equivalence relation?

3. Construct five different equivalence relations on $A = \{a,b,c,d\}$. Are the inverses of these relations also equivalence relations?

4. Is it true that the inverse of an equivalence relation is an equivalence relation?

5. In each of the following, a rule of correspondence for a relation $R = \{(a,b) \mid (a,b) \in Q \times Q\}$ is given. In each case, what would be the rule of correspondence for the inverse relation R^{-1}?

a) $b = a - 2$

b) $b = 3a + 6$

c) $b = a^2$

d) $b = \frac{7a - 5}{4}$

e) $b = 12 - 5a$

f) $b = \frac{7}{3} + \frac{5}{6}a$

Section 3
Functions

A particular type of relation, which is of immense importance in mathematics, is called a function.

Definition: A relation in which no two distinct ordered pairs have the same first coordinate is called a *function.*

Example 10: Let f = {(a,1), (b,1), (c,2), (d,2), (e,3)}. The ordered pairs of f are such that no two distinct ordered pairs of f have the same first coordinate. Hence, f is a function. The domain of f is the set {a,b,c,d,e} and the range of f is the set {1,2,3}.

We usually are concerned with a function from a set A to a set B.

Definition: A *function f from a set A to a set B* is a subset of $A \times B$ such that for each $a \in A$ there is a $b \in B$ such that $(a,b) \in f$. Symbolically, we write $f:A \to B$ and say "f is a function from A to B."

Notice that if $f:A \to B$, then $\mathfrak{D}(f) = A$ and $\mathfrak{R}(f) \subset B$.

Example 11: Let A = {1,2,3} and B = {a,b,c,d}. Then f = {(1,a), (2,d), (3,c)} is a function from A to B.

Recall that sometimes it was desirable to specify a relation by a rule of correspondence. Thus, we note that a function from a set A to a set B can be specified by a rule of correspondence that pairs each element of a set A with a unique element of set B.

Let f be a function from set A = $\{1,2,3\}$ to set B = $\{1,4,7,9\}$, and let the rule of correspondence be $b = 3a - 2$ where $a \in$ A and $b \in$ B. With this information, we can determine the set of ordered pairs $\{(1,1), (2,4), (3,7)\}$. Notice that each element of A is associated with exactly one element of B.

The following definition presents some frequently used notation and terminology associated with the notion of function.

Definition: Let f be a function from set A to set B, f:A → B. The element *b* in B that is associated with an element *a* in A under f is called the *image of a under f* or the *value of f at a*, denoted f(*a*) = *b*. Also, *a* is called a *pre-image of b under f.*

Notice that if $(a,b) \in$ f, where f is a function, then a is a pre-image of b under f and that b is the image of a under f. With this new notation we can represent a function f from a set A to a set B by f = $\{(a,\text{f}(a)) \mid a \in \text{A and f}(a) \in \text{B}\}$.

Example 12: Let A = {a,b,c} and B = {at,cat,bat,sat,hat, rat}. Let f be a function from set A to set B such that f = {(a,at), (b,bat), (c,cat)}. We can write f(a) = at, f(b) = bat, and f(c) = cat. We say that a is the pre-image of at under f, and at is the image of a under f. We can make analogous statements for b and bat, and c and cat.

It is sometimes convenient to consider the set of all images of some subset of the domain of a function f; that is, if f:A → B and C ⊂ A, then $\text{f(C)} = \{\text{f}(a) \mid a \in \text{C}\}$. f(C) is called the image set of set C under f.

Exercises 4.3

1. Which of the following are functions? Indicate the domain and range of those that are functions.

a) $\{(3,-1), (4,12), (5,-1), (6,-2)\}$

b) $\{(7,\text{r}), (7,\text{s}), (7,\text{t})\}$

c) {(happy, glad), (joy, glad), (glad, delighted)}

d) $\{(-1,7), (8,4), (5,-3), (0,-2), (-3,1), (-1,4)\}$
e) $\{(4,-2), (6,8), (3,7), (-13,12), (7,-2), (6,-12)\}$
f) $\{(a,b), (x,y), (p,q), (a,r), (s,b), (t,p)\}$

2. Which of the following specify functions?

a) x is the mother of y where x and y are elements of the set of all people.
b) x is the next-door neighbor of y where x and y are elements of the set of all people in Detroit.
c) $\{(x,y) \mid y = x^2$ where $x \in \{-1,0,1\}$ and $y \in Z\}$
d) $\{(x,y) \mid y = x$ where $x \in \{0,1,4,9\}$ and $y \in Z\}$
e) $\{(x,y) \mid y = (x + 3)(x - 3)$ where $x \in \{-2,-1,0,1,2\}$ and $y \in Z\}$

3. Let $f: Z \rightarrow Z$ be specified by the rule of correspondence $f(a) = a^2 + 5a - 1$.

a) What is the image of 5 under f?
b) What is the image of -1 under f?
c) What are the pre-images of 13 under f?
d) -6 is the pre-image of what number?
e) What is the value of f at 7?

4. Let f be a function from set X to set Y; that is, $f: X \rightarrow Y$. Let $A \subset X$ and $B \subset X$. Show that $f(A \cup B) = f(A) \cup f(B)$ by the set-inclusion method. Hint: We must show that:

1) $f(A \cup B) \subset f(A) \cup f(B)$
2) $f(A) \cup f(B) \subset f(A \cup B)$

1) Let $y \in f(A \cup B)$. This means $y = f(c)$ where $c \in (A \cup B)$. Then $c \in A$ or $c \in B$. So, $f(c) \in f(A)$ or $f(c) \in f(B)$. Thus, $f(c) \in f(A) \cup f(B)$. Hence, $y \in f(A) \cup f(B)$. Thus, $f(A \cup B) \subset f(A) \cup f(B)$.
2) You do this one.

5. Let f be a function from set X to set Y; that is, $f: X \rightarrow Y$. Let $A \subset X$ and $B \subset X$. Show that $f(A \cap B) \subset f(A) \cap f(B)$ by the set-inclusion method.

Section 4
Ways of Representing Functions

There are several ways of specifying functions. Let us consider the *squaring function* f from A = $\{-1, 1, 0, -2, 2\}$ to B = $\{0, 1, 2, 3, 4, 5\}$. This function can be represented by the following methods.

1. Listing the ordered pairs $\{(-1,1), (1,1), (0,0), (-2,4), (2,4)\}$

2. Expressing the rule of correspondence

a) We can specify f by the *statement* "the second coordinate of each ordered pair is the square of the first coordinate."

b) We can also specify by *formula*: $f = \{(a,b) \mid b = a^2$ where $a \in A$ and $b \in B\}$.

3. Using a table

The ordered pairs of f can be specified by a table.

a	b
−1	1
1	1
0	0
−2	4
2	4

4. Using a diagram

The following diagram completely determines the ordered pairs in f.

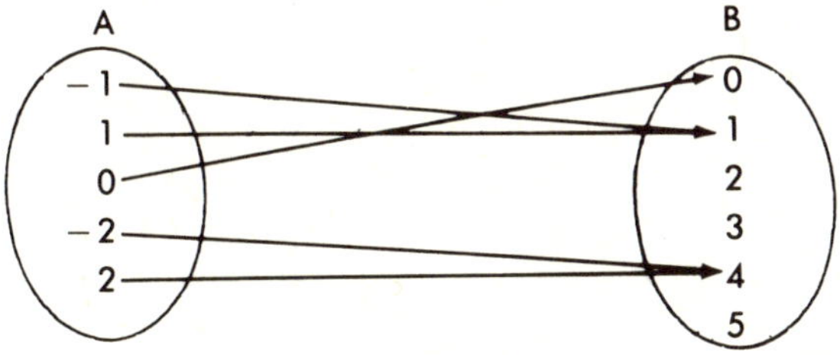

5. Using a coordinate diagram

We can represent the Cartesian product of the two sets A and B by a coordinate diagram in the following way. Draw two perpendicular lines, one horizontal and one vertical. Label the vertical line "B" and the horizontal line "A," and label the point of intersection "O". We will call this point of intersection the *origin*. Think of the elements of set B as being equally spaced from the origin upward (not including the origin) on the vertical line. Then, think of the elements of set A as being equally spaced to the right of the origin (not including the origin) on the horizontal line. Now, draw vertical lines parallel to vertical line B through the points on line A and draw horizontal lines parallel to horizontal line A through the points on line B. The intersections of these vertical and horizontal lines correspond to the ordered pairs of A × B. We locate the appropriate intersection for a given ordered pair by first finding the first coordinate of the ordered pair on the A line. Then, follow the vertical line that passes through that point until it is intersected by the horizontal line that passes through the second coordinate of the ordered pair found on line B. This intersection point corresponds to the given ordered pair. Thus, for sets A and B as given, we have the coordinate diagram shown in Figure 4.1. The points in the diagram corresponding to the ordered pairs of f are designated by heavy dots. The diagram in Figure 4.1 is called a *graph* of f.

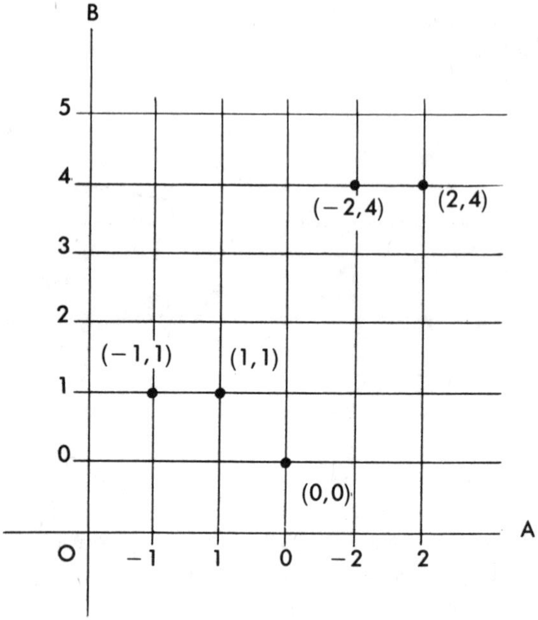

Figure 4.1

We have been able to express the squaring function f with domain A and range B in several different ways. We see that a function can be specified by a set of ordered pairs, a statement, a formula, a table, a diagram, and a graph. Not every function is conveniently expressed by all of these methods, but if you are aware of several ways of expressing functions, then you can choose the way that best suits your purpose.

Exercise 4.4

1. Which of the following diagrams specify a function from set A to set B?

a) 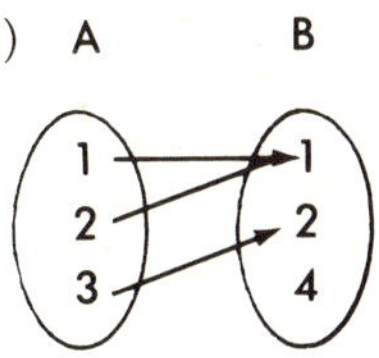

b) 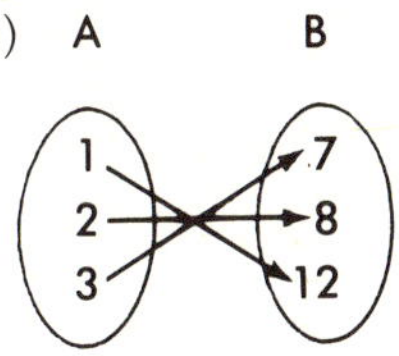

c) 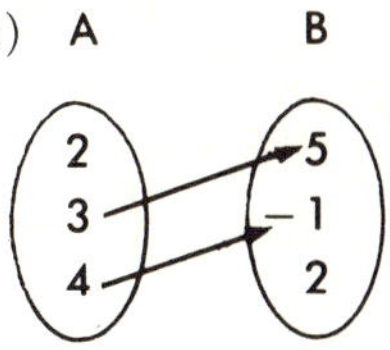

d) 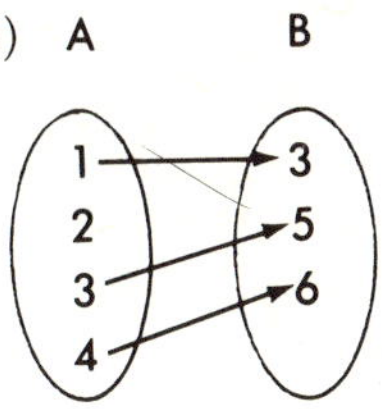

e) 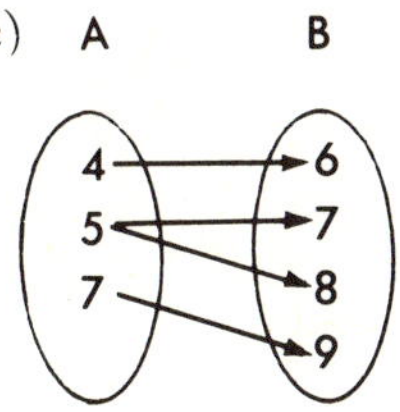

f) 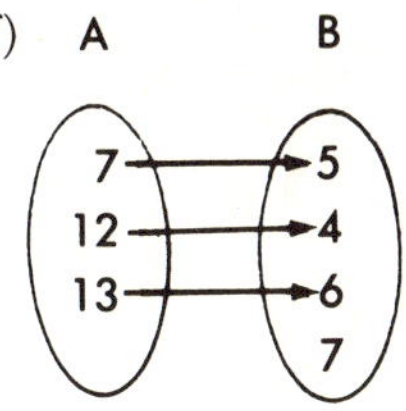

g) 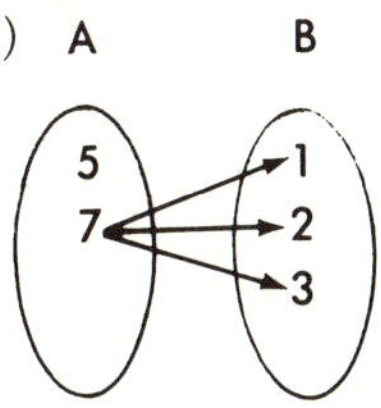

h) 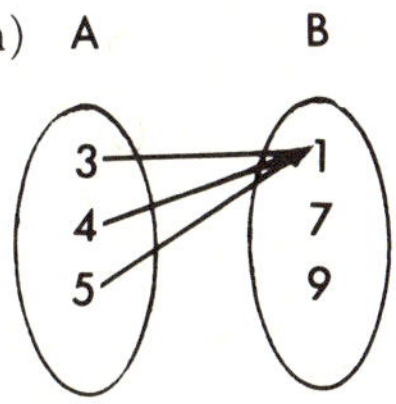

i) 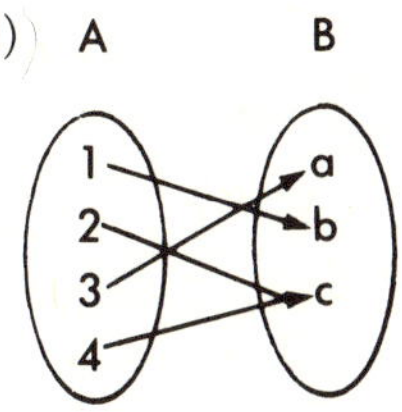

2. Let X = {a, b, c} and Y = {b, c, d, e}. Which of the following tables specify functions from X to Y?

a)

x	y
a	b
b	c
c	d

b)

x	y
a	e
b	d
c	c

c)

x	y
a	c
b	e
d	c

d)

x	y
a	b
b	d
a	c

e)

x	y
a	b
b	c
b	d
c	e

f)

x	y
a	e
b	e
c	e

g)

x	y
a	b
b	a
c	e
c	d

h)

x	y
a	d
b	e
c	b

i)

x	y
a	e
b	d

3. Represent each of the functions below by first using a table then a diagram.

a) $f = \{(a, f(a)) \mid f(a) = a^2 + 2a \text{ for } a \in \{1,2,3,\ldots,8\}\}$

b) $f = \{(a, f(a)) \mid f(a) = \frac{a}{3} \text{ for } a \in \{-1,0,2,4,5,-8\}\}$

c) $f = \{(12,7), (4,8), (3,6), (-1,7), (9,12), (13,4)\}$

d) $f = \{(x, f(x)) \mid f(x) = \frac{(x+1)(x-4)}{x+2} \text{ for } x \in \{-4,-3,0,3,4\}\}$

4. Use a coordinate diagram to represent each of the functions below.

a) f = {(a,b), (b,c), (c,d), (d,e)}

b) f = {(#,$), (@,¢), (¢,$), (*,%), (%,¢)}

c) f = {(a,#), (@,e), (d,b), (*,#)}

Section 5
The Cartesian Plane

We have discussed using a coordinate diagram to represent the Cartesian product of two sets. The Cartesian plane is a coordinate diagram that is of fundamental importance in mathematics.

Let the set **R** represent the set of real numbers. Then we can regard the Cartesian plane as a plane in which the points are coordinatized in such a way that there exists a one-to-one correspondence between $\mathbf{R} \times \mathbf{R}$ (the set of all ordered pairs of real numbers) and the set of points composing the plane. Figure 4.2 presents the standard way of picturing the Cartesian plane.

The points on line x are in 1-1 correspondence with the set of real numbers, as are the points on line y. The arrowheads show that the lines extend indefinitely in the indicated direction. The four quadrants of the Cartesian plane are indicated by Roman numerals.

For a graph of the function f defined by the set of ordered pairs $\{(-1,1), (1,1), (0,0), (-2,4), (2,4)\}$, used in the previous section, we indicate the points of the Cartesian plane that correspond to $f \subset \mathbf{R} \times \mathbf{R}$ in Figure 4.3. Notice that the domain of f is $\{0,1,-1,2,-2\}$, the range of f is $\{0,1,4\}$, and the rule of correspondence for f is $y = x^2$.

When the domain and range of a function are subsets of the real numbers, as is the case with the above function, the

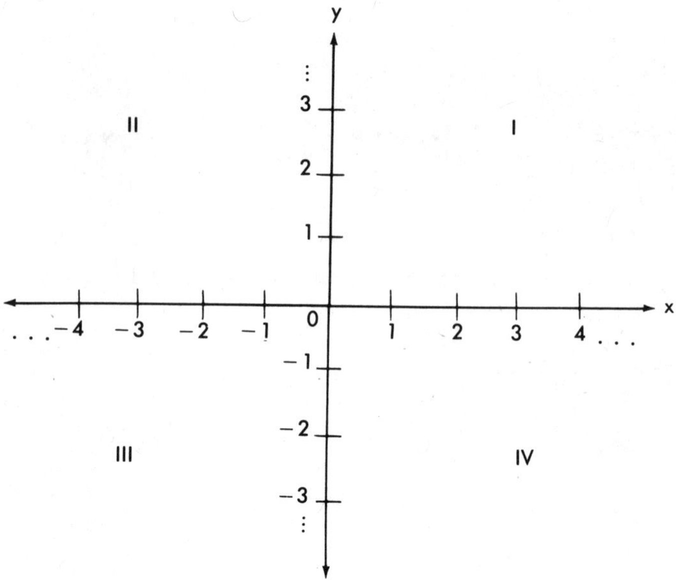

Figure 4.2

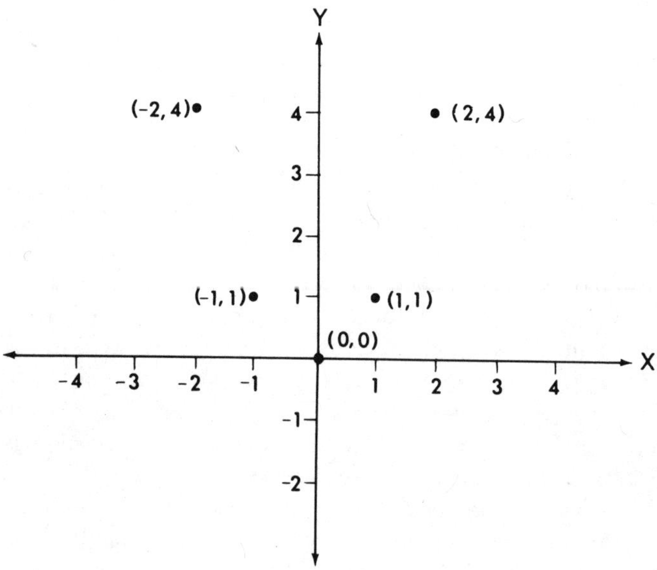

Figure 4.3

convention is to present the graph of the function in the Cartesian plane.

Consider the function f specified by the rule of correspondence $y = 2x + 1$, and suppose $\mathfrak{D}(f) = \mathbf{R}$. To sketch the graph for f, we first pick out a few points in $\mathfrak{D}(f)$ and find the corresponding value of f at these points (the y value; that is, $y = f(x)$). After recording these pairs in a table, we then locate the points in the Cartesian plane that correspond to the ordered pairs in the table. Since $\mathfrak{D}(f)$ consists of all values of **R**, we connect these points in the plane with a smooth curve (line) and indicate by arrowheads that this line is not the complete graph of f. Figure 4.4 shows a table and graph for f.

Since the ordered pairs of a function are such that no two distinct ordered pairs have the same first coordinate, the graph of a function never has more than one point on

X	Y
−2	−3
−1	−1
0	1
1	3
2	5
•	•
•	•
•	•

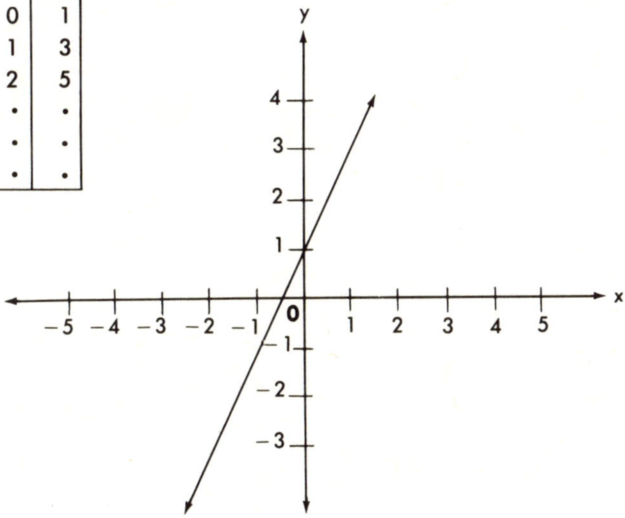

Figure 4.4

any vertical line. In the Cartesian plane, this means that if a vertical line intersects the graph of a function it intersects it in exactly one point.

This very brief discussion of the Cartesian plane and the graphs of functions in that plane does not begin to do justice to these topics. The applications of these concepts in mathematics and the sciences are extensive.

Exercises 4.5

1. Determine in which of the opposite graphs, if a vertical line intersects the graph, it intersects it in exactly one point.

2. Sketch the graph of each of the following functions in the Cartesian plane. Also, give a table specifying the function.

a) $\{(3,-2), (7,1), (5,-2), (0,0), (-3,2), (-4,-1)\}$
b) $\{(x,y) \mid y = x + 2 \text{ where } x,y \in \mathbf{R}\}$
c) $\{(x,y) \mid y = 3x + 4 \text{ where } x,y \in \mathbf{R}\}$
d) $\{(x,y) \mid y = 3x - 4 \text{ where } x,y \in \mathbf{R}\}$
e) $\{(x,y) \mid y = \frac{x+3}{2} \text{ where } x,y \in \mathbf{R}\}$

3. In which quadrant is each of the following points located?

a) $(2,3)$
b) $(-7,4)$
c) $(-2,-3)$

4. Functions are often referred to by other names. Sometimes functions are called *mappings* or *transformations*. These names give emphasis to certain intuitive notions associated with some functions. Using the Cartesian plane, graph the image of each of the following under the transformation $T:\mathbf{R} \times \mathbf{R} \to \mathbf{R} \times \mathbf{R}$ defined by $T(x,y) = (x + 2, y + 4)$. For example, $(2,3) \in \mathbf{R} \times \mathbf{R}$ is $(4,7)$ under the transformation T; that is, $T(2,3) = (4,7)$.

a) $(3,4)$
b) $(-2, 7)$
c) The x axis
d) The line $y = x$

a)
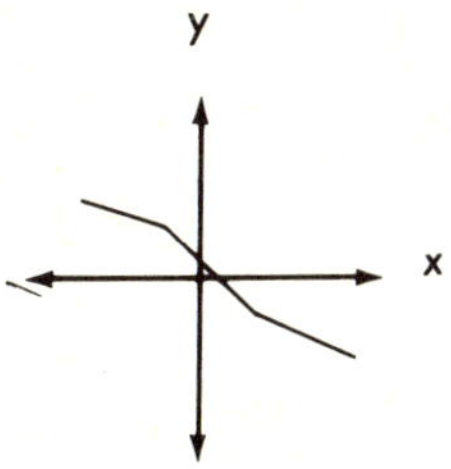

b)
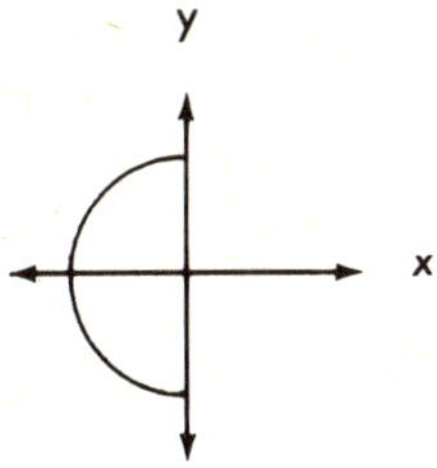

c)
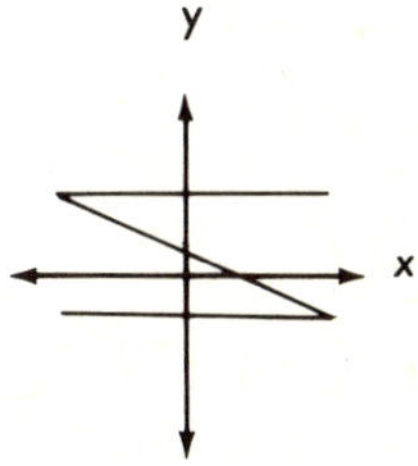

d)
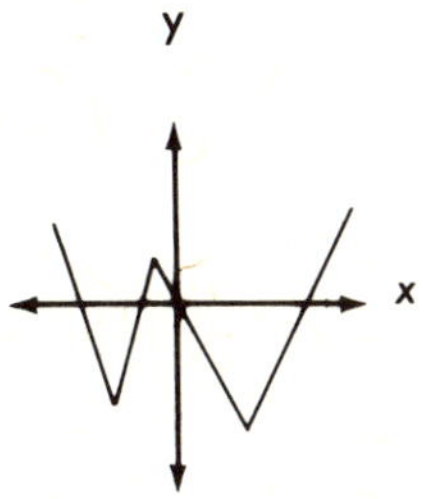

e)
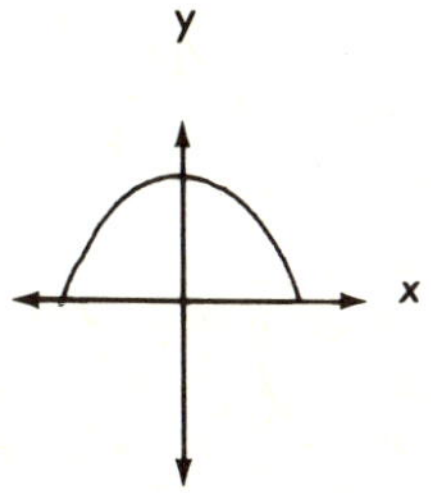

f)
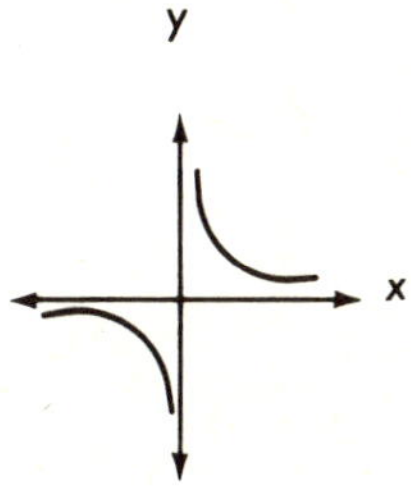

g)
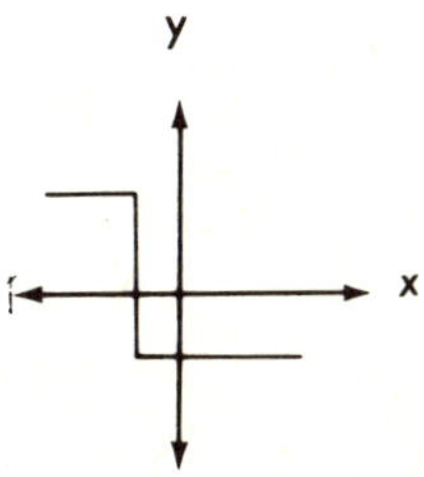

h)
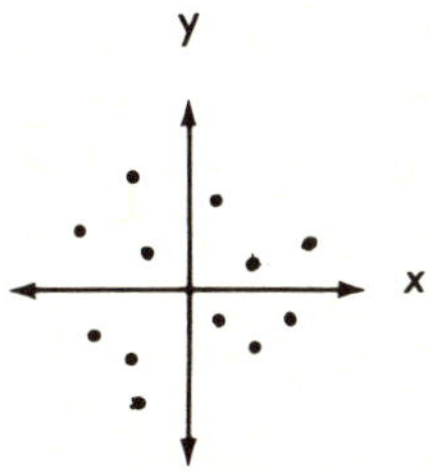

i)
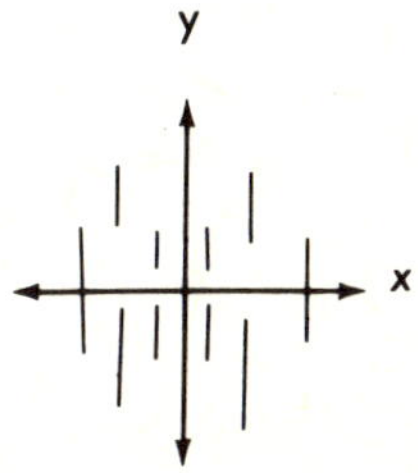

j)
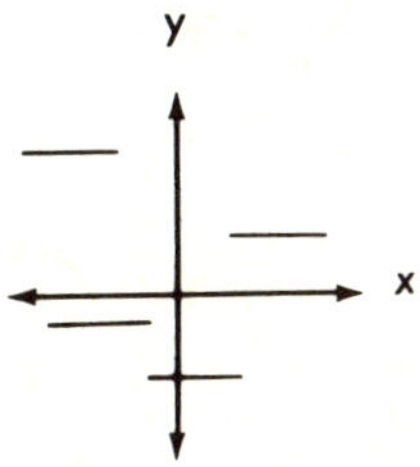

k)
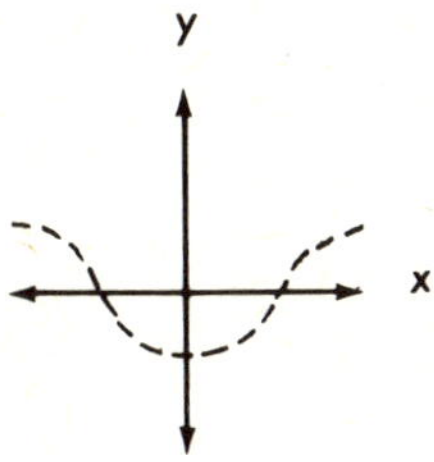

Section 6
Inverse of a Function

Since every function is a relation and the inverse of every relation is a relation, it is natural that we should consider the inverse of a function. Let f be a function. The inverse of the function f is its inverse as a relation. That is, $f^{-1} = \{(b,a) \mid (a,b) \in f\}$. The reader should realize that f^{-1} is not necessarily a function.

Example 13: The function f from set A to set B is specified by the diagram in Figure 4.5. Note, f = {(a,1), (b,1), (c,5), (d,5), (e,2), (f,3), (i,5)}. Suppose we reverse the arrows in Figure 4.5.

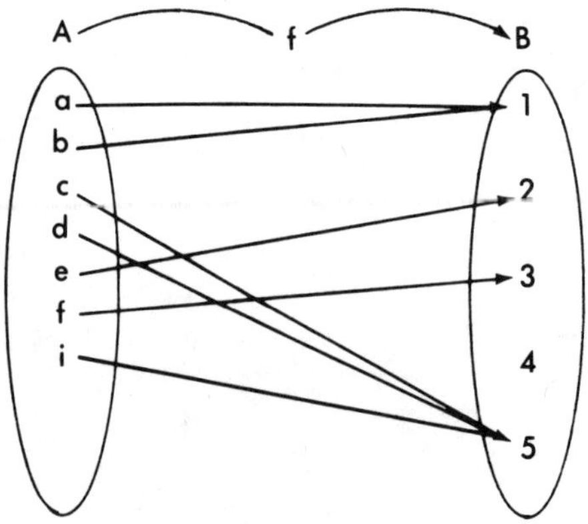

Figure 4.5

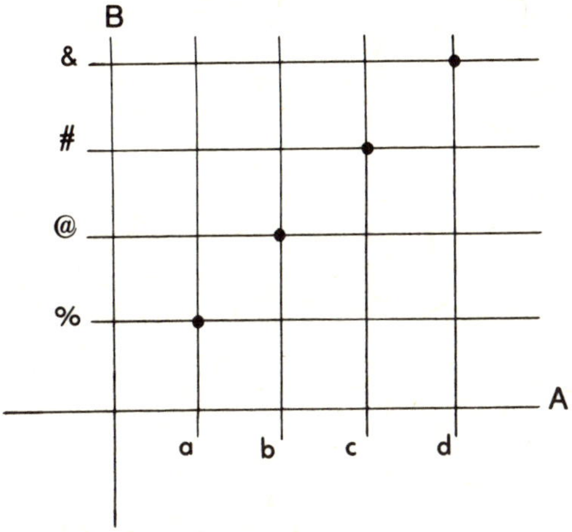

Figure 4.6

Then we get the associated set of ordered pairs {(1,a), (1,b), (2,e), (3,f), (5,c), (5,d), (5,i)}. Notice that this set of ordered pairs is the same set we get if we reverse the coordinates of the ordered pairs of f. Also note that f^{-1} = {(1,a), (1,b), (2,e), (3,f), (5,c), (5,d), (5,i)} is a relation that is not a function.

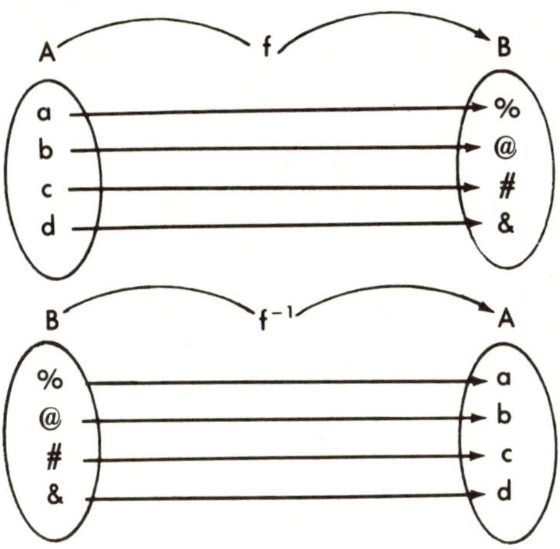

Figure 4.7

Example 14: Let f:A → B be specified by the graph in Figure 4.6. There are certain conditions under which we can be assured that f^{-1} is a function. In the next section, we will discuss these conditions. f and f^{-1} are shown by the diagrams in Figure 4.7. It is clear that f is a function from set A to set B, that f:A → B, and that f^{-1} is a function from set B to set A, f^{-1}:B → A.

Exercises 4.6

1. Using the Cartesian plane, draw the graph of each relation below and the graph of the inverse of each relation. Which of the relations are functions? Which of the inverse relations are functions? Which of the relations are functions whose inverses are functions?

a) $\{(1,5), (6,3), (-2,3), (-4,1), (1,2), (2,3)\}$
b) $\{(a,b) \mid b = a + 4 \text{ for } a \in \mathbf{R}\}$
c) $\{(a,b) \mid b = a^2 \text{ for } a \in \mathbf{R}\}$
d) $\{(3,4), (1,2), (-1,0), (0,3), (7,1), (-2,-3), (-3,3)\}$
e) $\{(a,b) \mid b = \frac{5}{a} \text{ for } a \in \{\ldots, -15, -10, -5, 5, 10, 15, \ldots\}\}$
f) $\{(a,b) \mid b = 3a \text{ for } a \in \mathbf{R}\}$
g) $\{(a,b) \mid b = a^2 + 3a + 5 \text{ for } a \in Z\}$
h) $\{(a,b) \mid b = 7a - 3 \text{ for } a \in \mathbf{R}\}$
i) $\{(a,b) \mid b = 2^a \text{ for } a \in Z\}$

2. Find the inverse relation for each of the following functions. Is f^{-1} a function?

a) $f = \{(3,5), (1,6), (5,-2), (6,1)\}$
b) $f = \{(x,y) \mid x + y = -1 \text{ where } x \in Z\}$
c) $f = \{(x,y) \mid y = x^2 - 9 \text{ where } x \in Z\}$
d) $f = \{(a,b) \mid a \in \{0, 1, 2\} \text{ and } b = a^2 + 4\}$
e) $f = \{(a,b) \mid a \in \{2, 4, 6, 8, 10\} \text{ and } b = a\}$

Section 7
Special Functions

Some types of functions occur so frequently in mathematical discourse that they have been given special designations.

Definition: A function f from set A to set B is called a *one-to-one function* if and only if $a = a'$ when (a,b) and $(a',b) \in$ f.

The following two statements are equivalent to the above definition of a one-to-one function. A function f:A → B is one-to-one if

a) different elements in B correspond to different elements in A; that is, $f(a) = f(a')$ implies $a = a'$.

b) no two different elements in A have the same image; that is, $a \neq a'$ implies $f(a) \neq f(a')$.

Example 15: Let f:A → B be a function specified by the diagram in Figure 4.8. Since different elements in B correspond to different elements in A, f is a one-to-one function.

Example 16: The function that associates the name of each county in Arizona with the name of its county seat is a one-to-one function since each county has exactly one county seat.

Definition: A function f from set A to set B is called an *onto* function if and only if the range of f equals B; that is, $\mathcal{R}(f) = B$.

Example 17: The diagram in Figure 4.9 specifies a function f. f:A → B is an onto function from set A to set B since the range of f equals B.

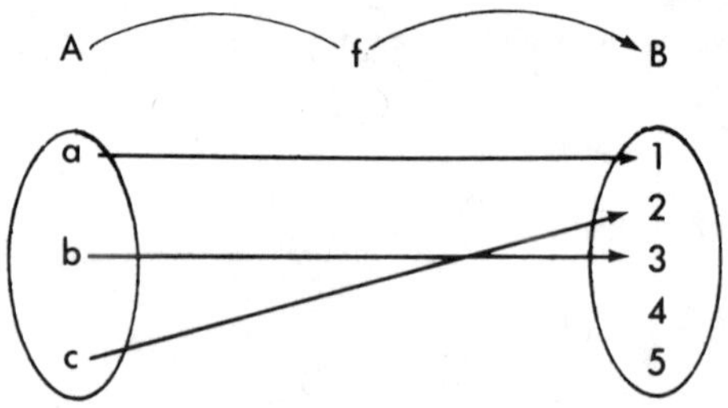

Figure 4.8

Example 18: Consider the function f from A = {1, 2, 3, 4} to B = {a} defined by {(1,a), (2,a), (3,a), (4,a)}. f is an onto function from A to B since $\mathcal{R}$(f) = B.

Example 19: Let A = {a, b, c} and B = {x, y, z}. Then, let f:A → B be specified by the graph in Figure 4.10. It is clear that f is an onto function from A to B.

In Chapter 1 of this book, we used some special types of functions, although we did not designate them as such. We said that two sets A and B were equivalent if a one-to-one correspondence existed between them. Now that we

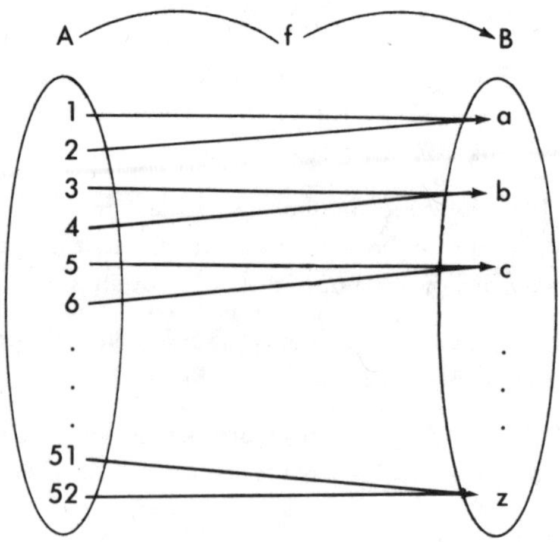

Figure 4.9

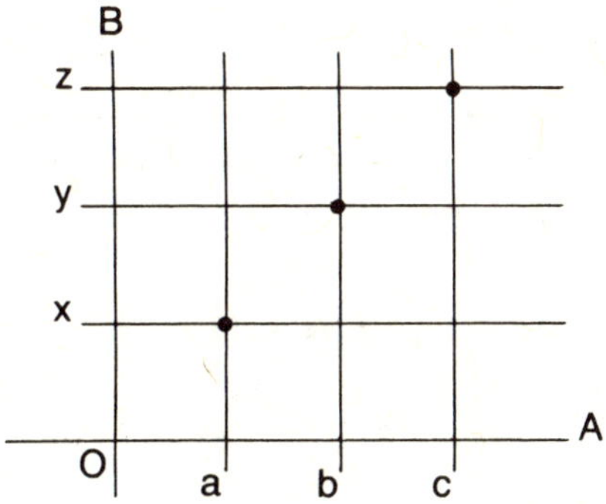

Figure 4.10

have the idea of function, we can cast the definition of a one-to-one correspondence into function language.

Definition: A *one-to-one correspondence* between sets A and B is a function f such that
1) f is one-to-one,
2) f is from A onto B.

A one-to-one correspondence is often referred to as a one-to-one onto function. When f is a one-to-one correspondence between sets A and B, the following statements are true.
a) If (a,b) and $(a,b') \in$ f, then $b = b'$, since f is a function.
b) If (a,b) and $(a',b) \in$ f, then $a = a'$, since f is one-to-one.
c) $\mathfrak{D}$(f) = A and $\mathfrak{R}$(f) = B, since f is from A to B and f is onto.
d) n(A) = n(B) and $n(\mathfrak{R}(f)) = n(\mathfrak{D}(f))$, since sets A and B are equivalent.

Example 20: The function f specified by the diagram in Figure 4.11 is a one-to-one correspondence. The diagram clearly illustrates the preceding remarks a) through d).

Example 21: The function f that associates each state in the United States with its capital is a one-to-one correspondence between states and their capitals.

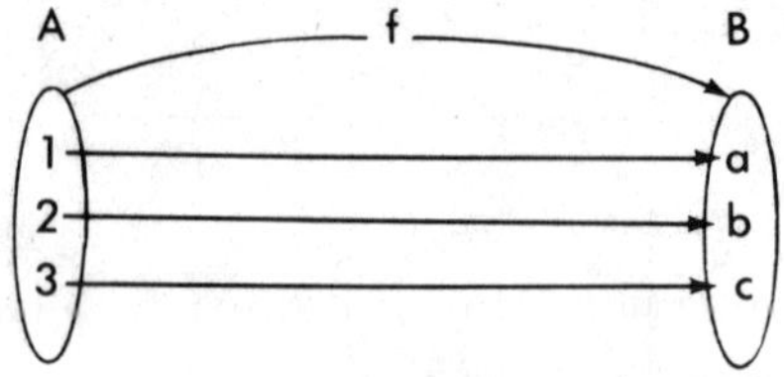

Figure 4.11

Definition: A function f from set A to set B is a *constant function* if and only if the range of f is a single element.

A constant function f:A → B is a set of ordered pairs such that the second coordinate of the ordered pairs is the same for each ordered pair in the function.

Example 22: The function f = {(a, 1), (b, 1), (c, 1), (d, 1)} is a constant function. $\mathfrak{D}(f) = \{a, b, c, d\}$ and $\mathfrak{R}(f) = \{1\}$.

Example 23: The diagram in Figure 4.12 specifies a constant function f from set A to set B.

Suppose we want to determine if two functions f and g are equal. Since f is a set of ordered pairs and g is a set of ordered pairs, it must be the case that f = g if and only if f is the same set of ordered pairs as g. This idea is embodied in the following definition.

Definition: Functions f and g are *equal functions* if and only if f and g are defined on the same domain A and for all $a \in A$, $f(a) = g(a)$.

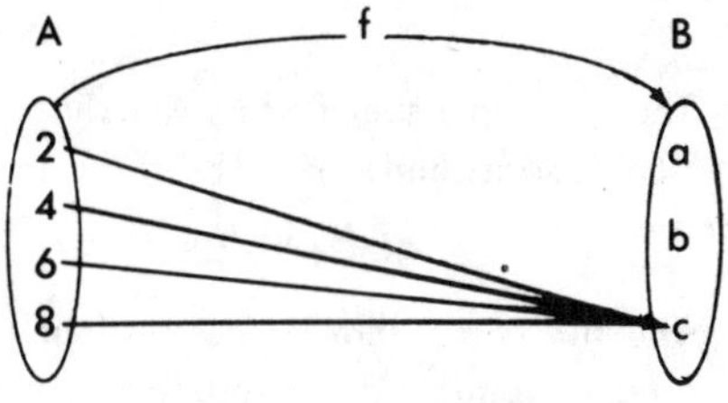

Figure 4.12

In using the definition of function equality to determine if two functions f and g are equal, we must check to see if $\mathfrak{D}(f) = \mathfrak{D}(g)$ and if $f(a) = g(a)$ for all a in their domains.

Example 24: Consider the function f specified in Figure 4.13. Consider the function g specified by the domain $A = \{-2, -1, 0, 1, 2\}$ and the rule of one-to-one correspondence $g(a) = a^2$. Now both f and g have the same domain, and f and g associate the same image with each element in their domains. Hence, $f = g$.

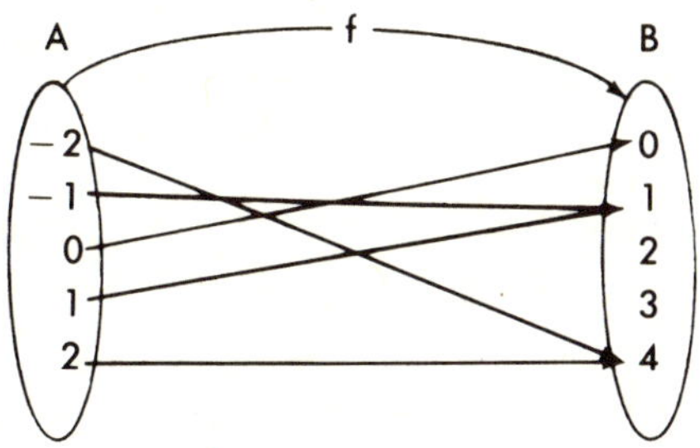

Figure 4.13

Example 25: Let $\mathfrak{D}(f) = A = \{0, 1\}$, and specify f by the rule of correspondence $f(a) = a + 1$. Let $\mathfrak{D}(g) = A = \{0, 1\}$, and specify g by the rule of correspondence $g(a) = 2a^2 - a + 1$. Now, $f(0) = 1 = g(0)$ and $f(1) = 2 = g(1)$. So, $f = g$. Also notice that $f = \{(0,1), (1,2)\} = g$.

Definition: Let f be a function on set A. f is the *identity* function on A if and only if $f = \{(a,a) \mid a \in A\}$.

From the above definition, we note that if $f:A \rightarrow A$ is the identity function on A, this means that for all $a \in A$, $f(a) = a$.

Example 26: Let f be the function on set A specified by the graph in Figure 4.14. Notice that for all $a \in A$, $f(a) = a$. That is, $f = \{(a,a), (b,b), (c,c), (d,d)\}$. Hence, f is the identity function on A.

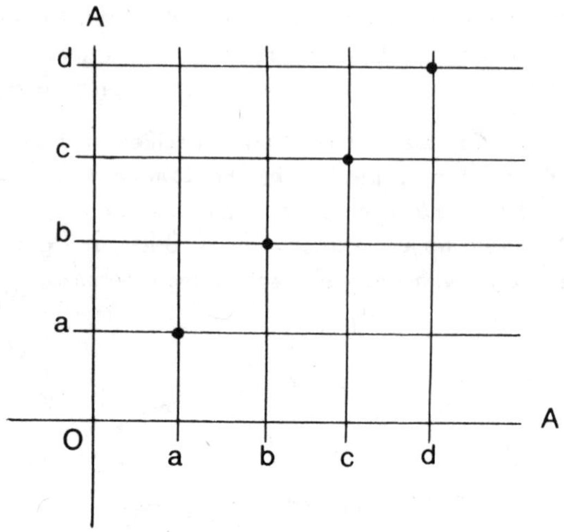

Figure 4.14

Example 27: Figure 4.15 specifies a function f on a set A. It is clear that f is the identity function on A.

In considering the inverse of a function f, we noted that f^{-1} is always a relation but not necessarily a function. Whether the inverse of a function is a function or not is of considerable interest in many mathematical settings.

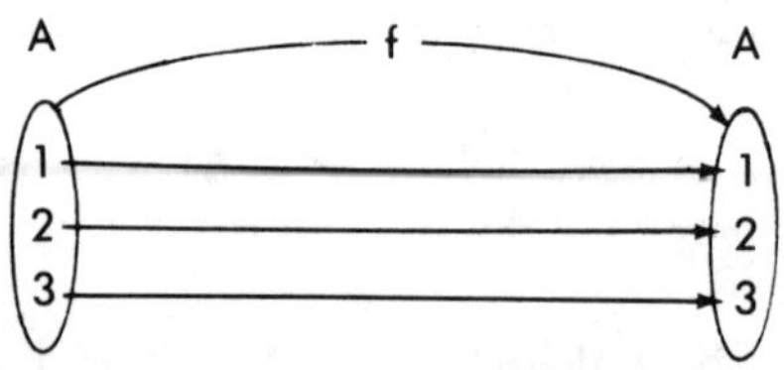

Figure 4.15

Definition: The inverse relation of a function is called an *inverse function* if and only if the inverse relation is a function.

Let f be a function from A to B. Now, if f is a one-to-one onto function, then for any $b \in B$ there will be an ordered pair of f such that b will be the second coordinate of the ordered pair and b will be associated with a unique $a \in A$. Hence, if f is a one-to-one onto function from A to B, then f^{-1} is a function from B to A, that is $f^{-1}: B \rightarrow A$.

For instance, let f be as specified in Figure 4.16. Now, f is a one-to-one onto function where f = {(*, a), (#, b), (@, c)}.

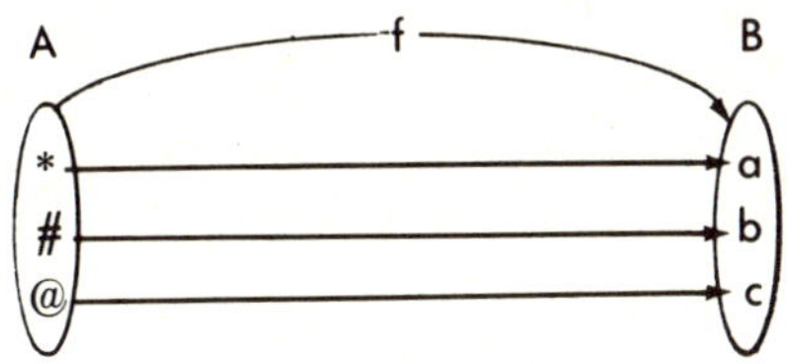

Figure 4.16

So, f^{-1} = {(a, *), (b, #), (c, @)} as depicted in the diagram in Figure 4.17. Notice that f^{-1} is the inverse function of f.

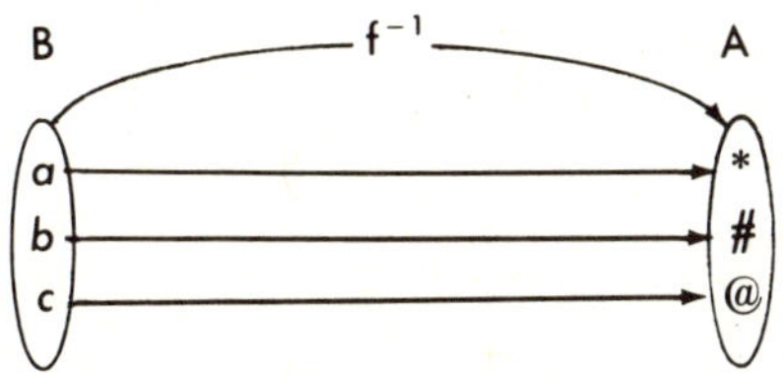

Figure 4.17

In summary, when $f: A \rightarrow B$ is one-to-one and onto, then f^{-1} is the inverse function of f.

Example 28: Let f be the function from A = {a, b, c} to B = {d, e, f} with ordered pairs {(a,d), (b,e), (c,f)}. Then f^{-1} = {(d,a), (e,b), (f,c)} is the inverse function of f. Notice that f is a one-to-one function from A onto B.

Example 29: The diagram in Figure 4.18 shows a function $f: A \rightarrow B$ and the inverse function $f^{-1}: B \rightarrow A$.

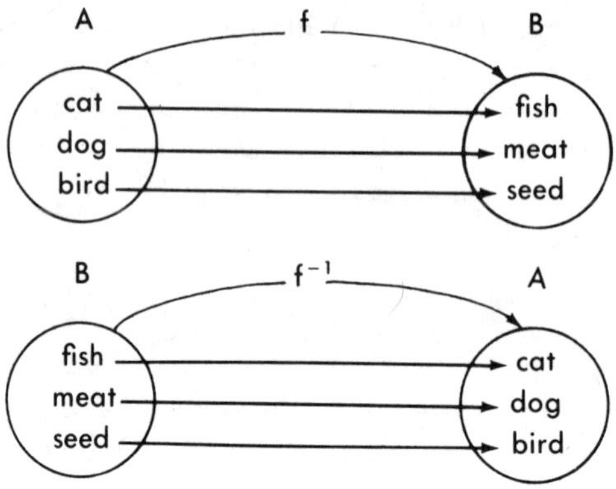

Figure 4.18

Exercises 4.7

1. Let A = {a, b, c, d}, B = {1, 2, 3} and C = {2}. Let f:A → C and g:A → B be defined as:

$$f = \{(a,2), (b,2), (c,2), (d,2)\}$$
$$g = \{(a,2), (b,2), (c,2), (d,2)\}$$

a) Is f an onto function?
b) Is g an onto function?
c) Is $\mathfrak{D}(f) = \mathfrak{D}(g)$?
d) Is f = g?
e) Is $\mathfrak{R}(f) = \mathfrak{R}(g)$?
f) Is B = $\mathfrak{R}(g)$?

2. Which of the diagrams in Problem 1 of Exercises 4.4 describe one-to-one onto functions?

3. How many functions are there from {a, b} to {a, b}? In each case, how many of these functions are onto? One-to-one?

4. Is the identity function always one-to-one and onto?

5. Let f be a function from set A to set B. Is it ever possible for $n(\mathfrak{D}(f))$ to be less than $n(B)$? For $n(\mathfrak{D}(f))$ to be less than $n(\mathfrak{R}(f))$?

6. Let f and g be specified by the following formulas. Is f = g?

$$f(x) = \frac{(x+2)(x-2)}{x+2},\ x \in \mathbf{R}, x \neq -2$$

$$g(x) = x - 2,\ x \in \mathbf{R}$$

7. Let f and g be specified by the following formulas. Is f = g?

$$f(x) = \frac{x - \frac{1}{3}}{x + \frac{2}{7}},\ x \in \mathbf{R},\ x \neq -\frac{2}{7}$$

$$g(x) = \frac{21x - 7}{21x + 6},\ x \in \mathbf{R},\ x \neq -\frac{2}{7}$$

8. Let A = {1, 2, 3, 4, . . .}, and let B = {5, 10, 15, 20, . . .}. We can establish a one-to-one correspondence between these two infinite sets as follows:

$$\begin{array}{lcccccc} A = \{1, & 2, & 3, & 4, & \ldots\} \\ \quad\ \updownarrow & \updownarrow & \updownarrow & \updownarrow & \\ B = \{5, & 10, & 15, & 20, & \ldots\} \end{array}$$

We can describe this same correspondence in function notation as follows: let $f: A \to B$ be defined by $f(n) = 5n$, for $n \in A$.

a) Show that f is a one-to-one function.
b) Show that f is an onto function.
c) Define f^{-1} by using function notation.

9. In each of the following, you are given two infinite sets. Determine a one-to-one correspondence between the two sets, and describe that correspondence in function notation.

a) A = {1, 2, 3, . . .} B = {3, 4, 5, . . .}
b) A = {2, 4, 6, 8, . . .} B = {1, 3, 5, 7, . . .}
c) A = {2, 6, 10, 14, . . .} B = {6, 14, 22, 30, . . .}

10. Let f be a function from set X to set Y; that is, $f:X \rightarrow Y$. Let $A \subset X$ and $B \subset X$. Show that if f is one-to-one, then $f(A \cap B) = f(A) \cap f(B)$. (Hint: In Exercises 4.3, Problem 5, you showed that $f(A \cap B) \subset f(A) \cap f(B)$ for any function so that all you must show is that $f(A) \cap f(B) \subset f(A \cap B)$.)

Section 8
Composition of Functions

Consider the functions f and g specified in Figures 4.19 and 4.20, respectively. The function f is from A to B and the function g is from B to C.

In Figure 4.21, we have again specified functions f and g. Notice that, by combining the diagrams for functions f and g, each element in set A corresponds with exactly one element in set C. If we let each element from set A be the first coordinate of an ordered pair and associate this element with the element in set C indicated by the arrows in Figure 4.21 as the second coordinate, then the set of ordered pairs {(1,*), (2,*), (3,#), (4,@)} is a function from A to C.

This idea of "hooking functions together" is very useful. The relationship that must exist between two functions f and

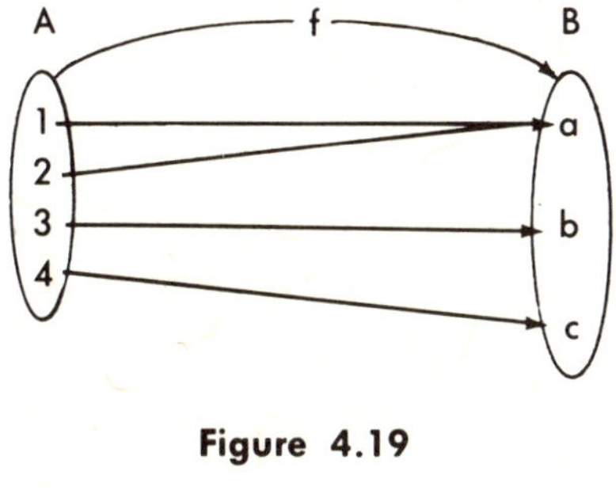

Figure 4.19

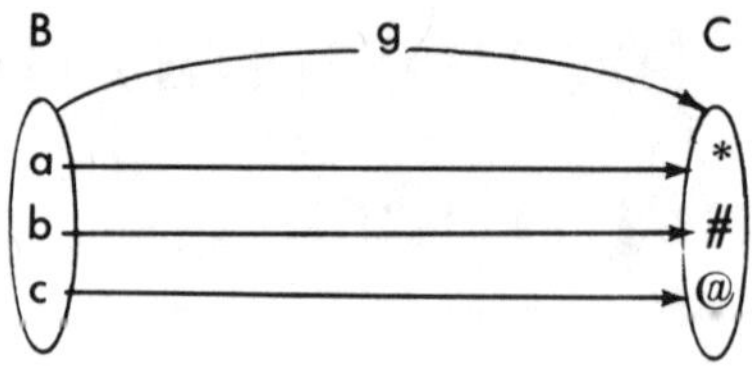

Figure 4.20

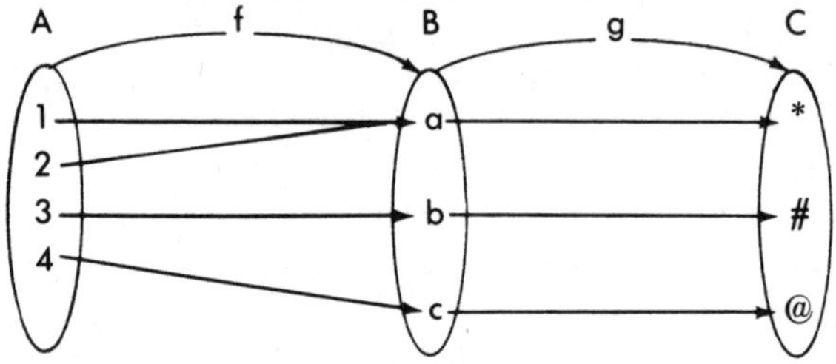

Figure 4.21

g before one can consider using them together is as follows: If f:A → B and g:C → D, then C must equal B.

Definition: Let f be a function from set A to set B and let g be a function from set B to set C. The *composite function of g and f,* denoted $g \circ f$, is the set $\{(a, g(f(a))) \mid a \in A\}$.

Notice that $g \circ f(a) = g(f(a))$ for $a \in A$. So, we can write $g \circ f = \{(a, g \circ f(a)) \mid a \in A\}$.

Example 30: Let functions f and g be designated according to the diagram in Figure 4.22.

We compute $g \circ f: A \to C$.

$$g \circ f\,(a) = g(f(a)) = g(1) = \gamma$$
$$g \circ f\,(b) = g(f(b)) = g(1) = \gamma$$
$$g \circ f\,(c) = g(f(c)) = g(3) = \alpha$$

Thus, $g \circ f = \{(a, \gamma), (b, \gamma), (c, \alpha).\}$

Example 31: Let f be a function from Z to Z specified by the rule of correspondence $f(a) = a^2$ for $a \in Z$. We specify the function g from Z to Z by the rule of correspondence $g(a) =$

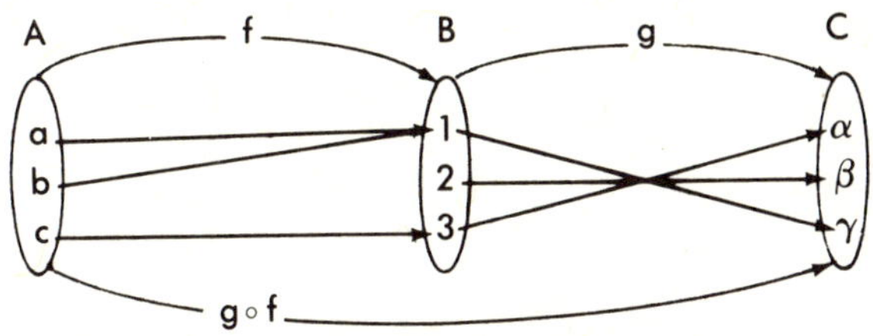

Figure 4.22

$a + 1$ for $a \in Z$. Notice that $f{:}Z \to Z$ and $g{:}Z \to Z$. Hence, $g \circ f{:}Z \to Z$. Values of $g \circ f$ at 1, 2, 3, and -4 are given below.

$$g \circ f(1) = g(f(1)) = g(1) = 2$$
$$g \circ f(2) = g(f(2)) = g(4) = 5$$
$$g \circ f(3) = g(f(3)) = g(9) = 10$$
$$g \circ f(-4) = g(f(-4)) = g(16) = 17$$

We can also consider $f \circ g$ since $f \circ g{:}\ Z \to Z$ is defined. The values of $f \circ g$ at 1, 2, 3, and -4 are:

$$(f \circ g)(1) = f(g(1)) = f(2) = 4$$
$$(f \circ g)(2) = f(g(2)) = f(3) = 9$$
$$(f \circ g)(3) = f(g(3)) = f(4) = 16$$
$$(f \circ g)(-4) = f(g(-4)) = f(-3) = 9$$

Notice that $g \circ f$ and $f \circ g$ are not the same function. This can also be illustrated by giving the rules of correspondence for $g \circ f$ and $f \circ g$.

$$(g \circ f)(a) = g(f(a)) = g(a^2) = a^2 + 1$$
$$(f \circ g)(a) = f(g(a)) = f(a + 1) =$$
$$(a + 1)(a + 1) = a^2 + 2a + 1$$

If $f{:}A \to B$ and $g{:}B \to C$, then $f \circ g$ may not exist (as with example 30 of this section), even though $g \circ f$ is defined. And, as example 31 of this section makes clear, $g \circ f$ and $f \circ g$ are not necessarily the same function, even if both of them are defined.

Consider the following functions: $f{:}A \to B$, $g{:}B \to C$, and $h{:}C \to D$. The domain and range of these functions are

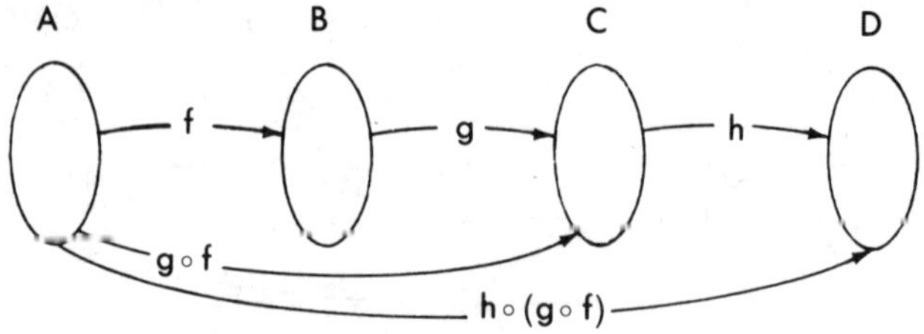

Figure 4.23

such that the functions $h \circ (g \circ f)$:A → D and $(h \circ g) \circ f$: A → D are defined. In Figure 4.23 we illustrate $g \circ f$:A → C and the function $h \circ (g \circ f)$:A → D. In a similar manner, the diagram in Figure 4.24 shows $h \circ g$:B → C and the function $(h \circ g) \circ f$:A → D. Since both $h \circ (g \circ f)$ and $(h \circ g) \circ f$ are functions from set A to set D, it is of interest to know if $h \circ (g \circ f)$ and $(h \circ g) \circ f$ are equal functions. Since both $h \circ (g \circ f)$ and $(h \circ g) \circ f$ have the same domain, all we need to do, according to the definition of equal functions, is to determine if $[h \circ (g \circ f)](a) = [(h \circ g) \circ f](a)$ for all $a \in A$. Now, for each $a \in A$,

$$[h \circ (g \circ f)](a) = h((g \circ f)(a)).$$

But, $(g \circ f)(a) = g(f(a))$ and thus we have $[h \circ (g \circ f)](a) = h(g(f(a)))$. Similarly, $[(h \circ g) \circ f](a) = (h \circ g)(f(a))$ but $(h \circ g)(f(a)) = h(g(f(a)))$. Hence, $[h \circ (g \circ f)](a) = h(g(f(a))) = [(h \circ g) \circ f](a)$ for all $a \in A$. We now have established that function composition is associative (i.e.,

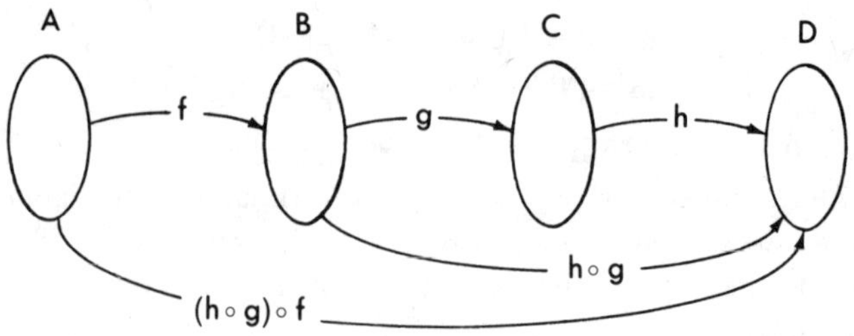

Figure 4.24

$h \circ (g \circ f) = (h \circ g) \circ f)$. This is an important result concerning function composition, as we will see in later work.

Exercises 4.8

1. Let f and g be specified by the following diagrams.

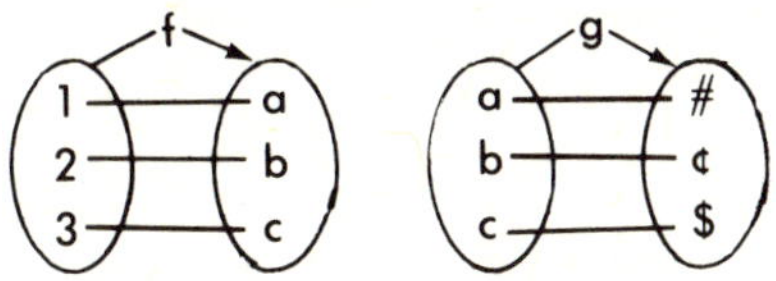

a) Draw a diagram that specifies $g \circ f$.
b) Write $g \circ f$ as a set of ordered pairs. Can you do this for $f \circ g$?

2. Let the functions $f:Z \to Z$ and $g:Z \to Z$ be specified by

$$f(x) = x^2 + 1 \quad \text{and} \quad g(x) = x + 2$$

Find:

a) $f \circ g\ (2)$
b) $g \circ f\ (2)$
c) $f \circ g\ (6)$
d) $g \circ f\ (6)$
e) $f \circ f\ (2)$
f) $g \circ g\ (2)$
g) $f \circ f\ (6)$
h) $g \circ g\ (6)$

3. Let the functions $f:Q \to Q$ and $g:Q \to Q$ be specified by

$$f(x) = \frac{4x + 2}{7} \quad \text{and} \quad g(x) = \frac{5x}{3}.$$

Find:

a) $f \circ g\ (-2)$
b) $g \circ f\ (4)$
c) $f \circ g \left(\frac{1}{3}\right)$
d) $g \circ f\ (7)$
e) $f \circ f\ (-3)$
f) $g \circ g\ (3)$
g) $f \circ f \left(\frac{1}{3}\right)$
h) $g \circ g\ (-7)$

4. Let the functions $f:Z \to Z$ and $g:Z \to Z$ be specified by

$$f(x) = 3x + 4 \quad \text{and} \quad g(x) = 5x - 2.$$

Find formulas that specify the rule of correspondence for:

a) $f \circ g$
b) $g \circ f$
c) $f \circ f$
d) $g \circ g$

5. Let the functions $f:Z \to Z$ and $g:Z \to Z$ be specified by

$$f(x) = x^2 - 2 \quad \text{and} \quad g(x) = x^2 - 1.$$

Find formulas that specify the rule of correspondence for the functions:

a) $f \circ g$
b) $g \circ f$
c) $f \circ f$
d) $g \circ g$

6. Show that, if $f:A \to B$ is onto and $g:B \to C$ is onto, then $g \circ f:A \to C$ is onto. (Hint: Let $c \in C$. Since g is onto, there exists a $b \in B$ such that $g(b) = c$. But, since f is onto, there exists an $a \in A$ such that $f(a) = b$. So, . . .).

7. Show that if $f:A \to B$ is one-to-one and $g:B \to C$ is one-to-one, then $g \circ f:A \to C$ is one-to-one.

8. Let functions f, g, and h be specified by the following diagrams.

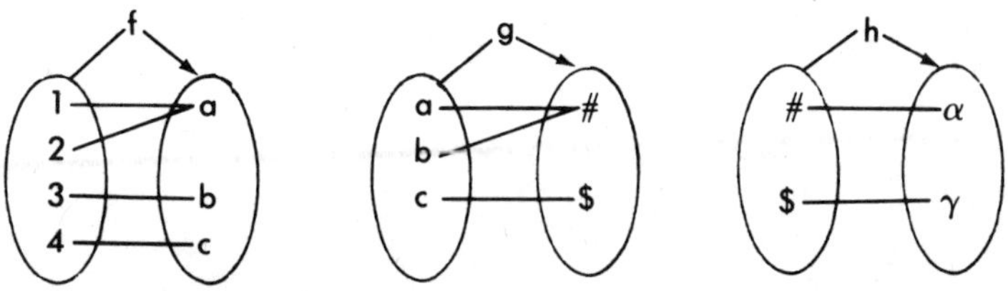

a) What is $[h \circ (g \circ f)](2)$?
b) What is $[(h \circ g) \circ f](2)$?
c) What function property does this illustrate?

Section 9

Further Considerations on the Inverse and Identity Function

Earlier, we noted that if f is a function from set A to set B such that f is one-to-one and onto, then f^{-1} exists. We now consider the composition of one-to-one and onto functions with their associated inverse functions.

Suppose $f: A \to B$ is one-to-one and onto. Then f^{-1}: $B \to A$ exists. Now, $f^{-1} \circ f$ is a function from set A to set A, as Figure 4.25 illustrates.

Let us follow an element a from set A to set A by way of the scenic tour $f^{-1} \circ f$ takes through set B. The image of a under f is $f(a)$, an element of B. But f^{-1} associates $f(a)$ with only a, since f is one-to-one. Hence, $f^{-1}(f(a)) = a$. That is, $f^{-1} \circ f = \{(a, f^{-1} \circ f\ (a)) \mid a \in A\}$. But, $f^{-1} \circ f\ (a) = a$ for each $a \in A$ means $f^{-1} \circ f = \{(a, a) \mid a \in A\}$, which is the

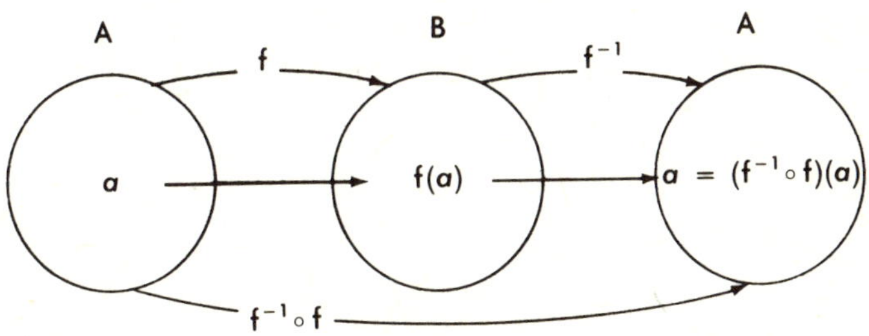

Figure 4.25

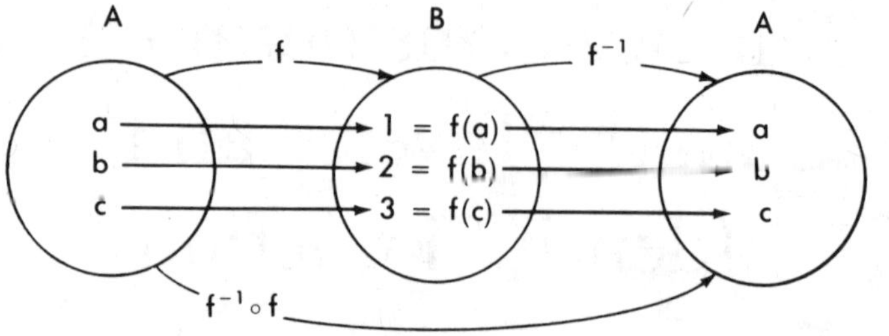

Figure 4.26

identity function on A. By similar consideration, we see that $f \circ f^{-1}: B \rightarrow B$ is the identity function on B.

Example 32: Let f be a function from set $A = \{a,b,c\}$ onto set $B = \{1,2,3\}$, and let f^{-1} be from set B onto set A, as specified in Figure 4.26. Now, $f^{-1}\circ f\ (a) = a$, $f^{-1}\circ f\ (b) = b$ and $f^{-1}\circ f\ (c) = c$, so that $f^{-1}\circ f = \{(a,a), (b,b), (c,c)\}$ and $f^{-1}\circ f$ is the identity function on A. (You should show that $f\circ f^{-1}$ is the identity function on B.)

Exercises 4.9

1. Let f and f^{-1} be as shown in Figure 4.26. Show that $f \circ f^{-1}$ is the identity function on B.

2. Let f be a function from set $A = \{3, 4, 5, 6\}$ onto set B, such that $f = \{(a, b) \mid b = 3a + 2 \text{ for } a \in A,\ b \in B\}$. Show by diagram and by tracing the elements of A that $f \circ f^{-1}$ is the identity function from B to B. Show that $f^{-1} \circ f$ is also the identity function from A to A.

Section 10
Binary Operations

Consider the set $N = \{1,2,3,\ldots\}$ and the set $N \times N = \{(1,1), (1,2), (1,3), \ldots, (2,1), (2,2), (2,3), \ldots, (3,1), (3,2), (3,3), \ldots\}$. Let f be specified by the rule of correspondence $f(a,b) = a + b$ for $a, b, \in N$. Now, if we add any two numbers in N, we get another number in N; that is, f is a function such that $f{:}N \times N \to N$. For instance, for $(2,3) \in N \times N$, $f(2,3) = 2 + 3 = 5$. Notice that we have cast the operation of addition on elements of N into function language. The domain of f is the set of ordered pairs in the Cartesian product $N \times N$, and the range of f is a subset of N. Functions whose domains are ordered pairs from the Cartesian product $A \times A$ of a set A and whose range is a subset of A are of special interest in mathematics.

Definition: Let A be a set. A function f from $A \times A$ to A is called a *binary operation* on A.

In defining a binary operation f on a set A, we must be sure that:

a) Each possible ordered pair in set $A \times A$ is associated with exactly one element of set A.

b) For each ordered pair in set $A \times A$, the element associated with it is again in set A. That is, A is *closed* under the binary operation f.

Example 33: Let $\mathbf{R}$ be the set of all real numbers. The following sets are binary operations of $\mathbf{R}$.

$A = \{((a,b), f(a,b)) \mid (a,b) \in \mathbf{R} \times \mathbf{R} \text{ and } f(a,b) = a \cdot b\}$

$B = \{((a,b), f(a,b)) \mid (a,b) \in \mathbf{R} \times \mathbf{R},$

$b \neq 0 \text{ and } f(a,b) = a/b\}$

$C = \{((a,b), f(a,b)) \mid (a,b) \in \mathbf{R} \times \mathbf{R} \text{ and } f(a,b) = a - b\}$

$D = \{((a,b), f(a,b)) \mid (a,b) \in \mathbf{R} \times \mathbf{R} \text{ and } f(a,b) = a + b\}$

We have written the four basic arithmetic operations in function notation; each function is a binary operation.

If the object of our work with binary operations was only to write the four basic arithmetic operations in function language, then the student might be justified in thinking that we complicated a "simple" notion. However, we will study other important binary operations besides the familiar binary operations of elementary arithmetic.

Example 34: Let A = {¢, &, $}.
a) Let $f = \{((a,b), f(a,b)) \mid (a,b) \in A \times A \text{ and } f(a,b) = ¢\}$
b) Let $f = \{((a,b), f(a,b)) \mid (a,b) \in A \times A \text{ and } f(a,b) = \$\}$
Both (a) and (b) are binary operations on A.

Example 35: Let A = {0, 1, 2, 4}, and define f as $\{((a,b), f(a,b)) \mid (a,b) \in A \times A$ and $f(a,b) =$ the minimum of a and $b\}$. Then, $f(1,2) = 1$, $f(2,2) = 2$, $f(0,4) = 0$, etc. f is a binary operation on set A.

Example 36: Sometimes a binary operation can be expressed by means of a table. Let A = {0, 1, 2}, and let f be specified by the table in Figure 4.27. $f(a,b) =$ the object in the intersection of the column headed by a and the row headed by b. Hence,

f	0	1	2
0	0	1	2
1	1	2	0
2	2	0	1

Figure 4.27

$f(1,2) = 0 =$ entry in the column headed by 1 and the row headed by 2. When specifying a binary operation f on a set A by a table, we list the elements across the top of the table in the same order as they are listed vertically.

When working with binary operations, we generally use the following notation. If $(a,b) \in A \times A$ then $f(a,b)$ will be written "$a * b$;" that is, $f(a,b) = a * b$ for all $(a,b) \in A \times A$. In general, we will use $*$ in this manner to denote a binary operation.

Example 37: Let Z be the set of integers. Then, $f = \{((a,b), f(a,b)) \mid (a,b) \in Z \times Z$ and $f(a,b) = a + b\}$ can be expressed with "$*$" as: $\{((a,b), a * b) \mid (a,b) \in Z \times Z$ and $a * b = a + b\}$.

We know that $a + b = b + a$ for all $a, b \in Z$. Hence, in example 37, $a * b = b * a$ for all $a, b \in Z$. Equivalently, the images of (a,b) and $(b,a) \in Z \times Z$ under f are the same; that is, $f(a,b) = f(b,a)$. The ability to recognize whether or not we get the same result when we change the order of two things is basic to our well-being in everyday activities. Suppose a man desires to shave, using an electric razor. It is important for him to know that he will have different results if he runs the razor over his face and then plugs it into an electrical outlet than he will have if he plugs it into an electrical outlet and then runs it over his face.

Definition: Let $*$ denote a binary operation on a set A, and let a and b be elements of A. The binary operation $*$ is *commutative* if and only if $a * b = b * a$ for all (a,b) and $(b,a) \in A \times A$.

We need to extend our discussion of binary operations so that we can consider an expression of the form $a * b * c$ for a, b, c elements of a set A with $*$ as a binary operation on set A. Obviously, we must group $a * b * c$ as $(a * b) * c$ or $a * (b * c)$ so that we can handle two elements of A at a time, as required by our definition of a binary operation. For most binary operations with which we will be working, we will be concerned with whether or not $(a * b) * c = a * (b * c)$.

Definition: Let $*$ be a binary operation on set A, and let a, b, $c \in A$. The binary operation $*$ is *associative* if and only if $(a * b) * c = a * (b * c)$ for all $(a * b, c)$ and $(a, b * c)$ element of $A \times A$.

Example 38: Let 2Z be the set of all even integers; that is, $2Z = \{\ldots, -6, -4, -2, 0, 2, 4, 6, \ldots\}$. Specify $*$ by $a * b = a + b$ for $(a,b) \in 2Z \times 2Z$. Clearly, $*$ is associative on 2Z.

Example 39: Consider the binary operation $*$ on Z specified by the rule of correspondence, "For $(a,b) \in Z \times Z$, $a * b = (a \cdot b) + 2$." This binary operation is *not* associative. Check to see that it is not. (Hint: $(a * b) * c = \{[(a \cdot b) + 2] \cdot c\} + 2$.)

Exercises 4.10

1. For each binary operation defined below, decide whether $*$ is commutative. Also, decide if $*$ is associative.

a) On Z, define $*$ by $a * b = a - b$

b) On **R**, define $*$ by $a * b = \dfrac{a}{b}$ where $b \neq 0$.

c) On Q, define $*$ by $a * b = 3^a \cdot 3^b$

d) On **R**, define $*$ by $a * b = \dfrac{3^a}{3^b}$

e) On **R**, define $*$ by $a * b = ab + 5$

f) On **R**, define $*$ by $a * b = a^b$

2. Which of the following are binary operations? Show by an example of each that is *not* a binary operation.

a) If $(x,y) \in Q \times Q$, then $x * y = \sqrt{x + y}$.

b) If $(x, y) \in Z \times Z$ and $y \neq 0$, then $x * y = \dfrac{x}{y}$

*	e	a	b	c
e	e	a	b	c
a	a	e	c	b
b	b	c	e	a
c	c	b	a	e

Figure 4.28

c) If $(x, y) \in N \times N$, then $x * y = x - y$.
d) If $(x, y) \in Z \times Z$, then $x * y = x^2 \cdot y^2$.

3. Let $*$ be an operation that is both associative and commutative. Show that $(a * b) * (c * d) = c * (b * d) * a$

4. Let A = {e, a, b, c}. Consider the table in Figure 4.28. Does this table describe a binary operation from A × A to A? If so, is the operation commutative? Is it associative?

Section 11

Important Examples of Binary Operations

Binary operations play an important role in the study of deductive mathematical systems. The basic arithmetic operations are examples of binary operations that we use in everyday life. We now wish to investigate some additional types of binary operations.

Finite Arithmetics

From everyday experience with the arithmetic operations, we use numbers from the infinite set **R**. Recall that the whole numbers, W, are a subset of the real numbers. Now, let us restrict our consideration to subsets of the whole numbers W. We will show a way to define a type of "addition" and "multiplication" on finite subsets of W.

Consider the set {0, 1, 2}. We will denote this set by Z_3. Now, the following rule of correspondence will specify the binary operation $\oplus$ from $Z_3 \times Z_3$ to Z_3. We will call this binary operation *addition modulo 3* for any two elements a and b in Z_3. The rule is:

a) $a \oplus b = a + b$, if and only if $a + b$ is less than 3.

b) $a \oplus b = (a + b) - 3$, if and only if $a + b$ is greater than or equal to 3.

We determine all possible ordered pairs of $Z_3 \times Z_3$ and associate with each ordered pair the appropriate element in

$(0,0) : 0 \oplus 0 = 0$	$(2,1) : 2 \oplus 1 = 0$
$(1,0) : 1 \oplus 0 = 1$	$(0,2) : 0 \oplus 2 = 2$
$(2,0) : 2 \oplus 0 = 2$	$(1,2) : 1 \oplus 2 = 0$
$(0,1) : 0 \oplus 1 = 1$	$(2,2) : 2 \oplus 2 = 1$
$(1,1) : 1 \oplus 1 = 2$	

Figure 4.29

Z_3, as specified by our rule of correspondence. Thus, we obtain the information in Figure 4.29.

From our computation, $\oplus$ is verified as a binary operation on Z_3. It is usually easier to keep track of the information in Figure 4.29 by putting it in table form, as in Figure 4.30. Notice that $\oplus$ in Z_3 is a commutative binary operation. That is, for $a, b \in Z_3$, $a \oplus b = b \oplus a$. Another way of saying this is to say that, in the table (Fig. 4.30), the entry in the column for a and the row for b (for $a, b \in Z_3$) is the same as the entry in the column for b and the row for a. This clearly means that a binary operation specified by a table is commutative if and only if the entries in the table are symmetric with respect to the diagonal from the lower right corner of the table to the upper left corner of the table, referred to as the *principal diagonal*.

To determine if $\oplus$ on Z_3 is associative, we must check to see if the 27 different possible expressions of the form $(a \oplus b) \oplus c = a \oplus (b \oplus c)$ for $a, b, c \in Z_3$ are true. This is a great deal of effort, but if we check the expression in all 27 cases, we will find that the expression is true in all 27 cases. Hence, $\oplus$ on Z_3 is associative.

Let $Z_m = \{0, 1, 2, \ldots, m - 1\}$ where m is a natural num-

$\oplus$	0	1	2
0	0	1	2
1	1	2	0
2	2	0	1

Table for addition modulo 3 in Z_3

Figure 4.30

ber. We can construct a table for addition modulo m in a manner similar to that discussed for addition modulo 3 by use of the following definition.

Definition: Let a and b be elements of Z_m. The binary operation of *additional modulo m*, denoted $a \oplus b$, is specified by the following rule of correspondence, which associates $(a, b) \in Z_m \times Z_m$ with $a \oplus b \in Z_m$. $a \oplus b$ is the remainder in the division of $a + b$ by m.

Although a rigorous explanation is not provided, we note that $\oplus$ is both commutative and associative in any set Z_m.

The next item of discussion is that of the operation of multiplication for finite sets. Let us use $Z_3 = \{0, 1, 2\}$ and the following rule of correspondence to specify the binary operation $\odot$ from $Z_3 \times Z_3$ to Z_3. We will call this binary operation *multiplication modulo 3* of any two elements a and b in Z_3. The rule is:

a) $a \odot b = a \cdot b$, if and only if $a \cdot b$ is less than 3.

b) $a \odot b = (a \cdot b) - 3$, if and only if $a \cdot b$ is greater than or equal to 3.

In Figure 4.31, all possible ordered pairs of $Z_3 \times Z_3$ are associated with the element in Z_3 specified by the rule of correspondence given above. Again, we can put the information

$(0,0) : 0 \odot 0 = 0$ $\quad$ $(2,1) : 2 \odot 1 = 2$
$(1,0) : 1 \odot 0 = 0$ $\quad$ $(0,2) : 0 \odot 2 = 0$
$(2,0) : 2 \odot 0 = 0$ $\quad$ $(1,2) : 1 \odot 2 = 2$
$(0,1) : 0 \odot 1 = 0$ $\quad$ $(2,2) : 2 \odot 2 = 1$
$(1,1) : 1 \odot 1 = 1$

Figure 4.31

in Figure 4.31 in a table. The table in Figure 4.32 is symmetric with respect to its principal diagonal, indicating that $\odot$ is commutative in Z_3. Again, to check associativity for $\odot$, we need to check to make sure the 27 possible cases of the expression $(a \odot b) \odot c = a \odot (b \odot c)$ for $a, b, c \in Z_3$ are true. It is the case that $\odot$ is associative in Z_3.

For $Z_m = \{0, 1, 2, \ldots m - 1\}$, where m is a natural number, we can construct a table for multiplication modulo m in

⟐	0	1	2
0	0	0	0
1	0	1	2
2	0	2	1

Table for multiplication modulo 3 in Z_3

Figure 4.32

a manner similar to that for ⟐ in Z_3 by use of the following definition.

Definition: Let a and b be elements of Z_m. The binary operation of *multiplication modulo m*, denoted a ⟐ b, which associates $(a, b) \in Z_m \times Z_m$ with a ⟐ $b \in Z_m$, is specified by the following rule of correspondence. a ⟐ b is the remainder in the division of $a \cdot b$ by m.

The binary operation ⟐ is both commutative and associative in Z_m.

Often, we will want to use the rule for multiplication modulo m in Z_m. However, we will want to consider only all the nonzero elements. This set we will designate $Z_m^* = Z_m - \{0\}$. The table in Figure 4.33 specifies ⟐ for Z_3^*. Notice

Z_3^*	1	2
1	1	2
2	2	1

Multiplication modulo 3 in Z_3^*

Figure 4.33

that ⟐ is a binary operation in Z_3^*. That is, each ordered pair $(a, b) \in Z_3^* \times Z_3^*$ is associated with a unique $c \in Z_3^*$.

We now consider $Z_4^* = \{1, 2, 3\}$, and build a table (Figure 4.34) according to the criterion specified in the definition for multiplication modulo m. Notice that multiplication modulo 4 in Z_4^* is not a binary operation, since $(2, 2) \in$

⊙	1	2	3
1	1	2	3
2	2	0	2
3	3	2	1

Multiplication modulo 4 in Z_4^*

Figure 4.34

$Z_4^* \times Z_4^*$ is associated with 0 and $0 \notin Z_4^*$. Hence, one must check to see if multiplication modulo m is defined in Z_m^*.

To make a statement regarding the conditions under which multiplication modulo m is defined in Z_m^*, the following terms are helpful. A natural number m is said to be *prime* if and only if it has exactly two distinct divisors, 1 and m. A *composite natural number m* is one that has more than two distinct divisors.

Multiplication modulo m is a binary operation in Z_m^* if and only if m is a prime. We note that the divisors of 3 are exactly 1 and 3. So 3 is a prime, and multiplication modulo 3 is a binary operation in Z_3^*. But the divisors of 4 are 1, 2, and 4. Therefore, 4 is composite; hence, multiplication modulo 4 is not a binary operation in Z_4^*.

Rigid Motions of Geometric Figures

You are familiar with some of the more common geometric figures, such as squares, rectangles, and triangles. In general, when we refer to rigid motion of a geometric figure, we mean a motion that, when applied to the figure, does not tear or distort the shape of the figure and leaves the figure in its same relative position in the plane. Our particular use of this concept of rigid motion will be related to its association with the symmetry properties of a geometric figure.

Consider rectangle ABCD in Figure 4.35. The lines h and v indicate the two lines of symmetry for the rectangle, and point r represents the intersection of these two lines. We can perform four rigid motions on rectangle ABCD. Note that the letters ABCD indicate the position of the ver-

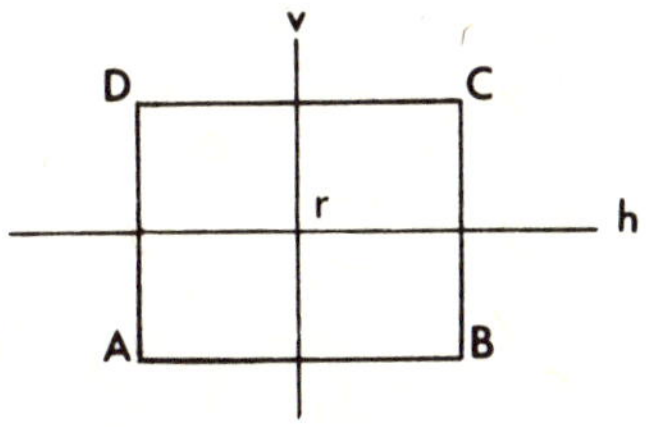

Figure 4.35

tices of the rectangle as they appear in the figure. Now suppose we "flip" the rectangle about line h in such a way that vertex B moves to where vertex C was and vertex C moves to where vertex B was. Similarly, D moves to A and A moves to D. We indicate this motion in a natural way by rectangular array H.

$$H = \begin{pmatrix} A & B & C & D \\ D & C & B & A \end{pmatrix}$$

Similarly, we can "flip" the rectangle about line v and describe this motion by the rectangular array

$$V = \begin{pmatrix} A & B & C & D \\ B & A & D & C \end{pmatrix}$$

Now, rotate the rectangle in the plane 180° about point r. Let this motion be denoted by R and specified by the array

$$R = \begin{pmatrix} A & B & C & D \\ C & D & A & B \end{pmatrix}$$

The fourth motion is the simplest. This is the motion that causes no change of vertices. If we denote this motion by I, then,

$$I = \begin{pmatrix} A & B & C & D \\ A & B & C & D \end{pmatrix}$$

With these four rigid motions of rectangle ABCD, we form the set $S = \{I, R, H, V\}$. It is natural to consider what the position of the vertices of the rectangle would be if we did one motion and then another. For instance, suppose we do R and then V. We see that motion R takes A to C and then motion V takes C to D, so that vertex A moves to where vertex D was originally. In a similar way, B goes to D and then D goes to C, so that vertex B moves to where vertex C was to begin with. Continue in a similar manner with C and D. We can specify the motion R and then V by the array

$$\begin{pmatrix} A & B & C & D \\ D & C & B & A \end{pmatrix}$$

Note that this array is the same as H. In Figure 4.36 we show the sequence for performing motion R and then motion V.

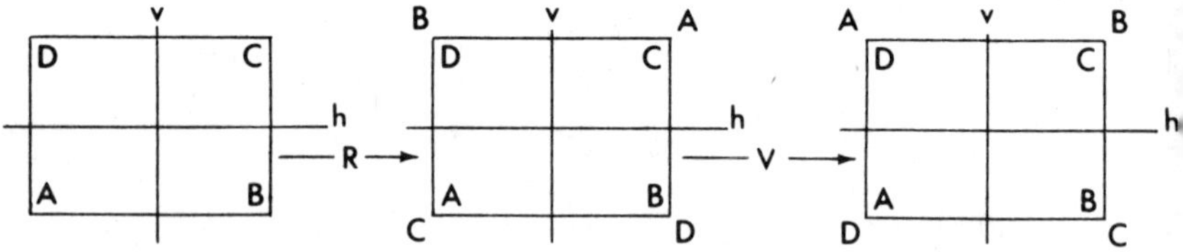

Original positions of vertices are inside rectangle.

Figure 4.36

We will denote the sequence of motions R and then V by V ◇ R. Notice that the symbol for the motion that is performed first is placed farthest to the right. This is done to keep this notation in harmony with our notation for the composition of functions. We can represent the sequence of motions R and then V by the array in Figure 4.37. The arrows indicate the path that vertex A takes on its scenic tour to vertex D's original position.

For binary combinations of motions in set S, we say: For x and y elements of S, $y \diamond x$ means to perform the mo-

$$V \diamond R = \begin{pmatrix} A & B & C & D \\ & & \downarrow & \\ B & A & D & C \end{pmatrix} \diamond \begin{pmatrix} A & B & C & D \\ \downarrow & & & \\ C & D & A & B \end{pmatrix} = \begin{pmatrix} A & B & C & D \\ \downarrow & & & \\ D & C & B & A \end{pmatrix} = H$$

Figure 4.37

tion x first and then follow it by the motion y. To keep track of all such binary combinations of motions, we build a table like the one in Figure 4.38. Note that $y \diamond x$ = the entry in the intersection of the column for the y motion and in the row for the x motion.

y replaced by these symbols

x replaced by these symbols

◇	I	R	H	V
I	I	R	H	V
R	R	I	V	H
H	H	V	I	R
V	V	H	R	I

Figure 4.38

Now let us consider some questions regarding the properties of the rigid motions of a rectangle. First, is $\diamond$ a binary operation? We note from the table in Figure 4.38 that $\diamond$ associates exactly one element in S with each ordered pair in S × S. Hence, $\diamond$ is a binary operation on S. Next, is $\diamond$ commutative? Again, by checking the table in Figure 4.38, we note that the table is symmetric about its principal diagonal. Hence, $\diamond$ is commutative. Is $\diamond$ associative? It is a long process to check this property, since we would need to check all 64 equations of the form $(z \diamond y) \diamond x = z \diamond (y \diamond x)$, where $x, y, z \in$ S. We will, however, verify associativity for a typical case.

$$(R \diamond V) \diamond H = H \diamond H = I$$
$$R \diamond (V \diamond H) = R \diamond R = I$$

Thus,

$$(R \diamond V) \diamond H = R \diamond (V \diamond H)$$

The operation $\diamond$, which we will call "and then," is an example of a binary operation on set S that is associated with the rigid motions of a rectangle. You will have opportunity to consider binary operations associated with the rigid motions of other geometric figures as you do the exercises for this section.

Permutations

Let set A = {a, b, c}. Consider all of the one-to-one functions from A to A which exist. These functions can be listed in the manner given in Figure 4.39.

$$f_1 = \begin{pmatrix} a \to a \\ b \to b \\ c \to c \end{pmatrix} \qquad f_2 = \begin{pmatrix} a \to b \\ b \to c \\ c \to a \end{pmatrix} \qquad f_3 = \begin{pmatrix} a \to c \\ b \to a \\ c \to b \end{pmatrix}$$

$$f_4 = \begin{pmatrix} a \to a \\ b \to c \\ c \to b \end{pmatrix} \qquad f_5 = \begin{pmatrix} a \to c \\ b \to b \\ c \to a \end{pmatrix} \qquad f_6 = \begin{pmatrix} a \to b \\ b \to a \\ c \to c \end{pmatrix}$$

Figure 4.39

Notice that each of the functions can be looked upon as specifying a rearrangement of the elements of set A. Thus, we say that each of these functions indicates a permutation of the elements of set A. It is convenient for us to think of permutations of elements of a set with *n* elements in terms of one-to-one onto functions or, equivalently, in terms of one-to-one correspondences.

Definition: Let A be a set with *n* elements. A *permutation* on set A is a one-to-one function from set A onto set A.

For the set A = {a, b, c} the one-to-one onto functions previously specified in Figure 4.39 are conveniently written in rectangular arrays, as in Figure 4.40.

$$f_1 = \begin{pmatrix} a & b & c \\ a & b & c \end{pmatrix} \qquad f_2 = \begin{pmatrix} a & b & c \\ b & c & a \end{pmatrix} \qquad f_3 = \begin{pmatrix} a & b & c \\ c & a & b \end{pmatrix}$$

$$f_4 = \begin{pmatrix} a & b & c \\ a & c & b \end{pmatrix} \qquad f_5 = \begin{pmatrix} a & b & c \\ c & b & a \end{pmatrix} \qquad f_6 = \begin{pmatrix} a & b & c \\ b & a & c \end{pmatrix}$$

Figure 4.40

We now consider the set $T = \{f_1, f_2, f_3, f_4, f_5, f_6\}$. Suppose we do two of the permutations in T in sequence. That is, suppose we do f_2 and then f_4. What is the result of such a sequence? This sequence is symbolized by $f_4 \circ f_2$ and means permutation f_2 is to be done first and then permutation f_4 should be done. Now,

$$f_4 \circ f_2 = \begin{pmatrix} a & b & c \\ a & c & b \end{pmatrix} \circ \begin{pmatrix} a & b & c \\ b & c & a \end{pmatrix}$$

f_2 matches a with b and f_4 matches b with c, so that f_2 and then f_4 matches a with c. Tracing b and c in a similar manner, we see that

$$f_4 \circ f_2 = \begin{pmatrix} a & b & c \\ c & b & a \end{pmatrix}$$

That is, another one-to-one correspondence has been constructed, namely f_5. Notice that it is natural to think of $f_4 \circ f_2$ in terms of function composition. That is,

$$f_4 \circ f_2 \ (a) = f_4(f_2(a)) = c$$
$$f_4 \circ f_2 \ (b) = f_4(f_2(b)) = b$$
$$f_4 \circ f_2 \ (c) = f_4(f_2(c)) = a$$

Let x and y be elements in the set T. Then $y \circ x$ means the composition of x by y. All possible combinations of the form $y \circ x$ where $x, y \in T$ are displayed in the table in Figure 4.41. The entry in the row headed by x and the column headed by y specifies $y \circ x$. We see that function

$\circ$	f_1	f_2	f_3	f_4	f_5	f_6
f_1	f_1	f_2	f_3	f_4	f_5	f_6
f_2	f_2	f_3	f_1	f_5	f_6	f_4
f_3	f_3	f_1	f_2	f_6	f_4	f_5
f_4	f_4	f_6	f_5	f_1	f_3	f_2
f_5	f_5	f_4	f_6	f_2	f_1	f_3
f_6	f_6	f_5	f_4	f_3	f_2	f_1

Figure 4.41

composition is a binary operation from T × T to T. Since function composition is associative, we know that $(z \circ y) \circ x = z \circ (y \circ x)$ for $x, y, z \in T$. We note that the binary operation of function composition is not commutative in T. That is, $y \circ x \neq x \circ y$ for all $x, y \in T$. In particular, $f_5 \circ f_2 \neq f_2 \circ f_5$. Notice also that the table in Figure 4.41 is not symmetric about its principal diagonal. Hence, we have an example of a noncommutative binary operation.

Exercises 4.11

1. List all the prime numbers from 1 to 50.

2. Give the following answers in Z_7.

a) $0 \oplus 6$ c) $4 \oplus 3$ e) $5 \oplus 2$
b) $6 \oplus 5$ d) $1 \oplus 5$ f) $3 \oplus 5$

3. Give the following answers in Z_{11}^*.

a) $1 \odot 10$ c) $10 \odot 10$ e) $7 \odot 9$
b) $3 \odot 8$ d) $9 \odot 9$ f) $5 \odot 3$

4. For which of the following is the binary operation $\odot$ defined?

a) modulo 8 in Z_8^* e) modulo 23 in Z_{23}^*
b) modulo 5 in Z_5 f) modulo 12 in Z_{12}
c) modulo 4 in Z_4 g) modulo 7 in Z_7
d) modulo 3 in Z_3^* h) modulo 16 in Z_{16}^*

5. Construct a table for addition modulo 6 in Z_6. Is addition modulo 6 commutative?

6. Construct a table for addition modulo 1 in Z_1.

7. Construct a table for multiplication modulo 6 in Z_6. Is multiplication modulo 6 in Z_6 a binary operation? Is it commutative?

8. Construct a table for multiplication modulo 5 in Z_5^*. Is this a binary operation?

9. List all of the rigid motions of the letter W. Construct a table showing all possible binary combinations of these rigid motions.

10. Use Figure 4.41 to verify each of the following.

a) $(f_3 \circ f_1) \circ f_2 = f_3 \circ (f_1 \circ f_2)$
b) $(f_4 \circ f_1) \circ f_5 = f_4 \circ (f_1 \circ f_5)$

c) $(f_6 \circ f_4) \circ f_3 = f_6 \circ (f_4 \circ f_3)$
d) $(f_5 \circ f_1) \circ f_4 = f_5 \circ (f_1 \circ f_4)$

11. Compute each of the following.

a) $\begin{pmatrix} 1 & 2 & 3 & 4 & 5 \\ 2 & 5 & 4 & 3 & 1 \end{pmatrix} \circ \begin{pmatrix} 1 & 2 & 3 & 4 & 5 \\ 3 & 5 & 1 & 2 & 4 \end{pmatrix}$

b) $\begin{pmatrix} 1 & 2 & 3 & 4 & 5 \\ 3 & 1 & 4 & 5 & 2 \end{pmatrix} \circ \begin{pmatrix} 1 & 2 & 3 & 4 & 5 \\ 5 & 4 & 3 & 2 & 1 \end{pmatrix}$

c) $\begin{pmatrix} 1 & 2 & 3 & 4 & 5 \\ 4 & 3 & 1 & 2 & 5 \end{pmatrix} \circ \begin{pmatrix} 1 & 2 & 3 & 4 & 5 \\ 5 & 4 & 1 & 2 & 3 \end{pmatrix}$

d) $\begin{pmatrix} 1 & 2 & 3 & 4 & 5 \\ 1 & 2 & 4 & 5 & 3 \end{pmatrix} \circ \begin{pmatrix} 1 & 2 & 3 & 4 & 5 \\ 2 & 1 & 4 & 5 & 3 \end{pmatrix}$

12. Compute each of the following.

a) $\left[\begin{pmatrix} 1 & 2 & 3 & 4 & 5 \\ 4 & 3 & 5 & 1 & 2 \end{pmatrix} \circ \begin{pmatrix} 1 & 2 & 3 & 4 & 5 \\ 2 & 4 & 1 & 5 & 3 \end{pmatrix}\right] \circ \begin{pmatrix} 1 & 2 & 3 & 4 & 5 \\ 3 & 4 & 2 & 5 & 1 \end{pmatrix}$

b) $\left[\begin{pmatrix} 1 & 2 & 3 & 4 & 5 \\ 4 & 3 & 5 & 1 & 2 \end{pmatrix} \circ \begin{pmatrix} 1 & 2 & 3 & 4 & 5 \\ 2 & 3 & 1 & 5 & 4 \end{pmatrix}\right] \circ \begin{pmatrix} 1 & 2 & 3 & 4 & 5 \\ 3 & 4 & 2 & 5 & 1 \end{pmatrix}$

c) $\begin{pmatrix} 1 & 2 & 3 & 4 & 5 \\ 3 & 1 & 2 & 5 & 4 \end{pmatrix} \circ \left[\begin{pmatrix} 1 & 2 & 3 & 4 & 5 \\ 2 & 1 & 5 & 4 & 3 \end{pmatrix} \circ \begin{pmatrix} 1 & 2 & 3 & 4 & 5 \\ 2 & 1 & 4 & 3 & 5 \end{pmatrix}\right]$

d) $\begin{pmatrix} 1 & 2 & 3 & 4 & 5 \\ 1 & 2 & 3 & 4 & 5 \end{pmatrix} \circ \left[\begin{pmatrix} 1 & 2 & 3 & 4 & 5 \\ 5 & 4 & 3 & 2 & 1 \end{pmatrix} \circ \begin{pmatrix} 1 & 2 & 3 & 4 & 5 \\ 5 & 4 & 3 & 2 & 1 \end{pmatrix}\right]$

13. Consider the equilateral triangle in the given figure. Let f_1 be a flip about line l, f_2 be a flip about line m, f_3 be a flip about line n, f_4 be a rotation about O of 120°, f_5 be a rotation about O of 240°, and f_6 be a rotation about O of 360°.

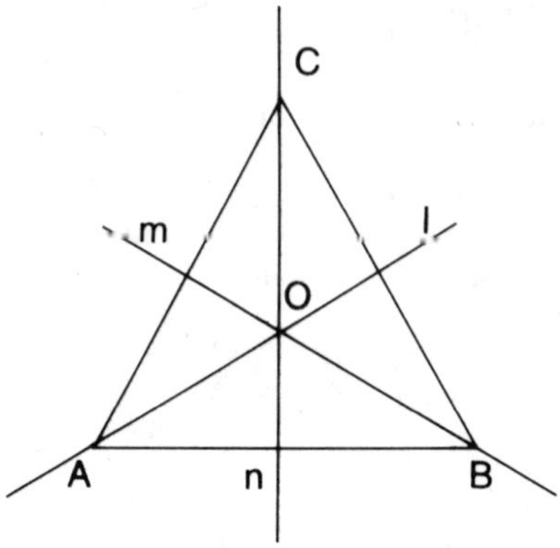

a) Indicate the effect each motion has on the vertices A, B, and C by use of rectangular arrays.

b) Construct a table for all possible binary combinations of motions on this triangle.

14. List all ten rigid motions of a regular pentagon. Using rectangular arrays, indicate the effect each motion has on the vertices of the pentagon.

15. List all eight rigid motions for a square. Construct a table showing all possible binary combinations of motions on the square.

Section 12
Concluding Remarks

In this chapter, you have been exposed to some of the more basic and useful concepts of function theory. Our discussion of functions has been greatly aided by terminology and concepts from set theory. Again we marvel at the continuity in mathematical theory that is made possible by set-theoretic concepts.

The basic arithmetic processes have provided us with examples of binary operations. These, as well as the other examples of binary operations in this chapter, will be important as we develop some interesting deductive mathematical systems in the next two chapters.

References Chapter Four

Fremont, Herbert, *How to Teach Mathematics in Secondary Schools* (Philadelphia: W. B. Saunders Company, 1969), pp. 172–202.

Heckenbach, Alan J., and Kegley, J. Colby, *Elementary Functions* (London: The Macmillan Company, 1970).

Laatsch, Richard, *Basic Algebraic Systems: An Introduction to Abstract Algebra* (San Francisco: McGraw-Hill Book Company, 1968), pp. 10–34.

Piagetian Cognitive-Development Research and Mathematical Education, proceedings of a conference conducted at Columbia University, October 1970 (Washington, D.C.: National Council of Teachers of Mathematics, 1971), pp. 12–33.

CHAPTER FIVE

Groups

We have considered several fundamental mathematical concepts. The study of sets, logic, deductive procedures, and functions has provided us with terminology and notation to facilitate greatly your work in deductive mathematical systems. Your labors will be rewarded as you see the vast area of mathematical thought resulting from the interweaving of the concepts we have discussed.

In this chapter, we will consider a deductive mathematical system of great importance to much of modern mathematics, to related fields such as physics and engineering, and, more recently, to the social sciences. The important abstract system to which we refer is called a *group*. The basic properties of a group are properties possessed by many seemingly unrelated branches of mathematics. A group can be represented in a variety of ways, in representations called *models*. We will examine a few of these models in this chapter. A group is one of the simplest deductive mathematical systems, yet group theory increases the power and beauty of mathematics immeasurably.

Section 1
Definition of a Group

Each of you has probably had the basic mathematical experience of solving simple equations. (An equation is an expression of equality that involves the use of the equals sign.) For example, the equation $x + 2 = 5$ is easily solved by any of us. Because we can recognize the solution to this equation so easily, we probably do not realize that fundamental mathematical properties must exist on the set of integers under the binary operation of addition before the equation can be solved.

The following example will help us to decide what the essential properties of a set A and a binary operation $*$ on A would need to be to solve any equation of the form $a * x = b$ for $a, b \in$ A. Recall the set T $= \{f_1, f_2, f_3, f_4, f_5, f_6\}$, which we discussed in the previous chapter, and the binary operation of function composition on T, as specified in Figure 4.11.

In the equation $f_2 \circ x = f_5$, we need to find some element of T with which to replace x so that this expression of equality will be true. We will consider the process of solving $f_2 \circ x = f_5$ to be that of isolating x on the left side of the equals sign. We will use the basic notion that what we do to one side of an expression of equality, we must do to the other side. This process of isolating x determines the right side of the equation, which is an expression that equals x. To see if this expression for x is a solution of the given equation, we replace x in the equation with this expression and see if a true statement results. The procedure shown in Figure 5.1 produces a possible replacement for x in the equation $f_2 \circ x = f_5$. Notice that to find a replacement value for x in the equation

$f_2 \circ x = f_5$	
$f_3 \circ (f_2 \circ x) = f_3 \circ f_5$	Compose both sides of this equation with f_3, which is the inverse function of f_2 in T.
$(f_3 \circ f_2) \circ x = f_3 \circ f_5$	We can do this step, since function composition is associative.
$f_1 \circ x = f_3 \circ f_5$	A function composed with its inverse function is the identity function f_1 in T.
$x = f_3 \circ f_5$	f_1 composed with any element of T is that element.
$x = f_6$	From table in Figure 4.41.

$f_2 \circ f_6 = f_5$ is true, so f_6 is a solution of the equation $f_2 \circ x = f_5$.

Figure 5.1

$f_2 \circ x = f_5$, the inverse function of f_2 (namely f_3) had to be an element of T. Also, the identity function f_1 was essential to the solution, as was the associative property. Since function composition is not commutative, it may be that $x \circ f_2 = f_5$ will not have the same solution as $f_2 \circ x = f_5$. However, to solve $x \circ f_2 = f_5$ through steps similar to those in Figure 5.1, we would need the identity function f_1, the inverse function of f_2 (namely f_3), and the associative property.

Generalizing the preceding discussion, it is apparent that, to solve these equations in T with the binary operation $\circ$, there must exist an element f_1 in T such that $f_1 \circ x = x \circ f_1 = x$ for all x in T. Also, for each x in T, there must exist a y in T such that $x \circ y = y \circ x = f_1$. The associative property of $\circ$ is also essential.

A deductive mathematical system in which we are always able to solve equations of the form $a * x = b$ or $x * a = b$ where a and b are elements of some set A and $*$ is a binary operation on A is of fundamental importance. A group is such a system.

Definition: A nonempty set G together with a binary operation $*$ on G is called a *group*, denoted $(G, *)$, if $*$ satisfies the following conditions.

G_1: The binary operation $*$ is associative on G.

G_2: There is an element e in G with the property that $a * e = e * a = a$ for all a in G. This element e is an *identity element* in G for the binary operation $*$.

G_3: For each element a in G, there is an element a' in G such that $a * a' = a' * a = e$ where e is the identity element. The element a' is an *inverse* of a relative to the binary operation $*$.

Definition: The *order of a group* is $n(G)$. If $n(G) = b$, such that b is a finite cardinal number, then $(G, *)$ is a group of finite order b. If b is an infinite cardinal number, then $(G, *)$ is a group of infinite order.

To help you get an idea of the great variety of representations or models for groups, we provide the following examples.

Example 1: Consider the set Z of integers. Under the binary operation of addition, Z is a group. That is,
1) Addition is associative.
2) There exists an element 0 such that for all $y \in Z$, $y + 0 = 0 + y = y$.
3) For each $y \in Z$, there is a y' $(-y)$ such that $y + y' = y' + y = 0$.

This is an example of a group of infinite order.

Now consider the binary operation of multiplication on Z. We see that for all $x \in Z$, $1 \cdot x = x \cdot 1 = x$. However, it is not the case that for each $y \in Z$ there exists $y' \in Z$ such that $y \cdot y' = y' \cdot y = 1$. Hence, Z is not a group under multiplication.

Example 2: The set Q of rational numbers under addition is a group. Also the set Q − {0} under multiplication is a group. (Q, +) and (Q − {0}, ·) are groups of infinite order.

Example 3: Consider the set $T = \{f_1, f_2, f_3, f_4, f_5, f_6\}$ and the binary operation on T of function composition. The following computation shows that every element of T has an inverse.

$$f_1 \circ f_1 = f_1 \circ f_1 = f_1$$
$$f_2 \circ f_3 = f_3 \circ f_2 = f_1$$
$$f_3 \circ f_2 = f_2 \circ f_3 = f_1$$
$$f_4 \circ f_4 = f_4 \circ f_4 = f_1$$
$$f_5 \circ f_5 = f_5 \circ f_5 = f_1$$
$$f_6 \circ f_6 = f_6 \circ f_6 = f_1$$

Also, T has an identity element under $\circ$, since:

$$f_1 \circ f_1 = f_1 \circ f_1 = f_1$$
$$f_2 \circ f_1 = f_1 \circ f_2 = f_2$$
$$f_3 \circ f_1 = f_1 \circ f_3 = f_3$$
$$f_4 \circ f_1 = f_1 \circ f_4 = f_4$$
$$f_5 \circ f_1 = f_1 \circ f_5 = f_5$$
$$f_6 \circ f_1 = f_1 \circ f_6 = f_6$$

We have already shown that function composition is associative. Hence, T under $\circ$ is a group of order six.

In Figure 5.2, we again present the table specifying the binary operation of function composition on T. Notice that there

$\circ$	f_1	f_2	f_3	f_4	f_5	f_6
f_1	f_1	f_2	f_3	f_4	f_5	f_6
f_2	f_2	f_3	f_1	f_5	f_6	f_4
f_3	f_3	f_1	f_2	f_6	f_4	f_5
f_4	f_4	f_6	f_5	f_1	f_3	f_2
f_5	f_5	f_4	f_6	f_2	f_1	f_3
f_6	f_6	f_5	f_4	f_3	f_2	f_1

Figure 5.2

is a horizontal row in the table identical to the top guide row and that there is a vertical column in the table identical to the left guide column. This circumstance indicates the presence of an identity element relative to ∘. Also, notice that f_1 appears in each row and each column. This indicates that every element of T has an inverse element relative to ∘. Additionally, in each row and each column each element of T appears exactly once. (The reason for this property is presented in Exercise 5.2, Problem 3.)

In general, if a table for a binary operation on a finite set is such that there is a row and a column exactly like the top guide row and left guide column, and if each element of the set appears exactly once in each row and in each column, then that structure is a group, provided the associative property holds.

Example 4: Think of the set S = {I, R, H, V} of rigid motions of a rectangle discussed in Chapter 4. In Figure 5.3, we again present the binary operation ◇ on S.

Notice that this table has the properties previously mentioned, thus indicating that (S, ◇) is a group. The order of (S, ◇) is four; that is, $n(S) = 4$.

◇	I	R	H	V
I	I	R	H	V
R	R	I	V	H
H	H	V	I	R
V	V	H	R	I

Figure 5.3

Example 5: Finite arithmetic structures afford us some other examples of groups. Consider the set Z_3 = {0, 1, 2} and the binary operation ⊕. By examining the table in Figure 5.4, we see that (Z_3, ⊕) is a group of order three.

Let us consider the set Z_3^* = {1, 2} under ⊙. The table in Figure 5.5 indicates that (Z_3^*, ⊙) is also a group. We point out that (Z_m, ⊕) is a group, but (Z_m^*, ⊙) is a group if and only if m is a prime.

$\oplus$	0	1	2
0	0	1	2
1	1	2	0
2	2	0	1

Figure 5.4

$\odot$	1	2
1	1	2
2	2	1

Figure 5.5

Example 6: Let U be a universal set and A and B be subsets of U. The symmetric difference of A and B (Exercise 1.4, Problem 24) is the set A △ B defined by

$$A \triangle B = (A \cap \bar{B}) \cup (\bar{A} \cap B).$$

Notice that:

1) △ is a binary operation defined on $\mathcal{P}(U)$. That is, for any two subsets A and B of U, $(A, B) \in \mathcal{P}(U) \times \mathcal{P}(U)$ and that A △ B is an element of $\mathcal{P}(U)$.

2) $(A \triangle B) \triangle C = A \triangle (B \triangle C)$ for A, B, and $C \in \mathcal{P}(U)$. (You should check this, even though the check is lengthy.)

3) $A \triangle \phi = \phi \triangle A = A$ for all $A \in \mathcal{P}(U)$; that is, ϕ is the identity element in $\mathcal{P}(U)$ under the operation △.

4) $A \triangle A = \phi$ for all $A \in \mathcal{P}(U)$; that is, the inverse of $A \in \mathcal{P}(U)$ relative to △ is A itself.

From the preceding steps, it is clear that $(\mathcal{P}(U), \triangle)$ is a group.

From the above examples, you should begin to appreciate the fundamental nature of the deductive mathematical system called a group because the basic properties of a group are possessed by each of the different models presented. This experience illustrates one of the merits of considering abstract formulations of mathematical properties.

A table specifying a binary operation $*$ on a set A that is symmetric about the principal diagonal indicates that $*$ is commutative. It happens that for many groups $*$ is commutative.

Definition: A group $(G, *)$ is *commutative*, or *Abelian*,* if and only if its binary operation $*$ is commutative.

*The word "Abelian" is used in honor of Niels Henrik Abel (1802–1829), who studied groups and made many significant contributions to mathematics in his brief lifetime.

*	e	a
e	e	a
a	a	e

Figure 5.6

Example 7: Let G = {e,a} and let * be specified by the table in Figure 5.6. (G,*) is a commutative group, since * is commutative.

Example 8: Since addition is commutative, (Z, +) is a commutative group.

Example 9: $(Z_m, \oplus)$ is a commutative group.

Example 10: In example four of this chapter (rigid motions of a rectangle), we have another illustration of a commutative group.

Example 11: Let U be a universal set, and let $\triangle$ be the symmetric difference of two sets that are elements of $\mathcal{P}(U)$. $(\mathcal{P}(U), \triangle)$ is a commutative group.

Example 12: Since function composition on $T = \{f_1, f_2, \ldots, f_6\}$ is not commutative, (T,o) is a noncommutative group.

Exercises 5.1

1. Consider the set A = {#, &} and the binary operation * on A given by the table.

*	#	&
#	#	&
&	&	#

Does A together with * form a group?

2. Consider the set of rigid motions of a square (Exercise 4.11, Problem 15.) Does the set of rigid motions of a square together with the operation "and then," denoted $\diamondsuit$, form a group?

3. Which of the following sets, together with the indicated operations, are groups?

a) $(3Z, +)$ for $3Z = \{\ldots, -6, -3, 0, 3, 6, \ldots\}$
b) $(Z, \cdot)$
c) $(\{0\}, +)$
d) $(\{1\}, +)$
e) $(\{1\}, \cdot)$
f) $(Z_6, \oplus)$
g) $(Z_6, \odot)$
h) $(Z_6^*, \odot)$
i) $(Z_5, \oplus)$
j) $(Z_5, \odot)$
k) $(Q - \{0\}, \cdot)$
l) $(\{x \in Z \mid x \text{ is odd}\}, +)$
m) $(\{x \in Z \mid x \text{ is odd}\}, \cdot)$

4. Let $*$ be defined on Z by $a * b = ab + 1$ for all $a, b \in Z$. Is $(Z, *)$ a group?

5. Let $*$ be defined on Z by $a * b = 2(a + b)$ for all $a, b \in Z$. Is $(Z, *)$ a group?

6. Consider the regular pentagon of Exercise 4.11, Problem 14. Does this set form a group under the operation of "and then"?

7. Decide if the following tables specify groups. State reasons for your answers.

a)

*	a	b
a	a	a
b	a	a

b)

*	1	−1	i	−i
1	1	−1	i	−i
−1	−1	1	−i	i
i	i	−i	−1	1
−i	−i	i	1	−1

c)

*	e	f	g
e	e	f	e
f	g	e	g
g	e	f	e

d)

*	#	@	$	&
#	#	@	$	&
@	@	#	&	$
$	$	&	#	@
&	&	$	@	#

8. Consider the figure  . Determine the rigid motions for this figure, and determine if the set of these rigid motions together with the binary operation "and then" is a group.

9. We specify the functions f and g by the following diagrams.

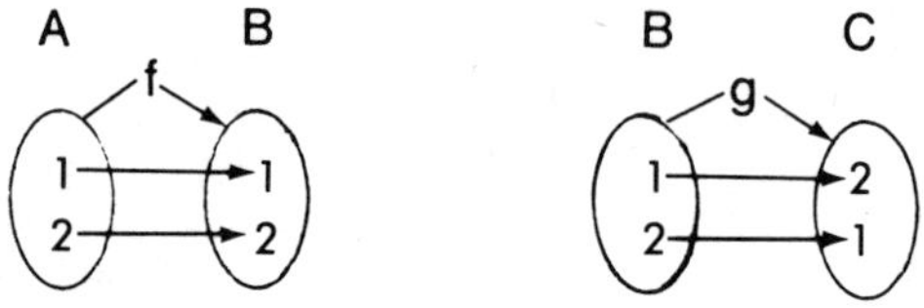

Let W = {f, g}. Does the set W together with the binary operation of function composition form a group?

10. For the set S = {I, R, H, V} and the binary operation $\diamondsuit$, does the equation $R \diamondsuit x = V$ have a solution? Let Y and Z be any elements of S. Does the equation $Y \diamondsuit x = Z$ have a solution? List all of the solutions of each of the following equations.

a) $x \diamondsuit x = I$

b) $x \diamondsuit x \diamondsuit x = I$

c) $y \diamondsuit x = R$

11. Consider the set $\{U, A, \bar{A}, \phi\}$ under the binary operation of $\cap$. Does $\{U, A, \bar{A}, \phi\}$ together with intersection form a group?

Section 2

Groups as Deductive Mathematical Systems

You have probably been wondering when our work would again make use of the deductive process. You can relax now and sharpen your pencil for the time has arrived.

In the definition of a group, clues exist indicating that the basic requirements for a deductive system are present. Undefined terms are used in the definition—the terms "set" and "operation," for instance. The properties of the binary operation on the abstract set G are actually assumptions or axioms. With these basic properties, we will be able to prove some statements for this system.

First, we need to make clear what we mean when we use the relation "is equal to."

Definition: Let a and b be elements of a set G. Element a is identical to element b, denoted $a = b$, if and only if a and b are the same object in the set G.

From this definition we see that if $a = b$ and c is any element of G, then $a * c = b * c$. It is also clear that if $a = b$ and $b = c$, then $a = c$.

We have noted undefined terms, definitions, and axioms pertaining to the deductive mathematical system $(G, *)$, where G is a set and $*$ is a binary operation on that set. Now all we need are statements that are candidates for theorems in our system. There are several questions you might ask yourself with respect to groups. Many of these questions could lead naturally to statements of the form $p \rightarrow q$. For instance, we might conjure up some statements as candidates for theorems by considering some properties of a group with which we are familiar, say $(Z, +)$. We know from experience that if $a + b = c + b$ for $a, b, c \in Z$ then $a = c$. Is this true for groups in general? That is, is it true that: If $(G, *)$ is a group and $a * b = c * b$ for a, b, c arbitrary elements of G, then $a = c$? We know from our past work in deductive logic that this statement is a theorem for this system if we can produce a sequence of statements about the objects in G under $*$ such that each statement is justified by:

1) a premise
2) a definition
3) an axiom
4) a previously proved statement (which we do not have as yet)
5) a principle of logic

Enough discussion! Let's do a proof. The format for our deductive process will be similar to the format we have used in previous work.

Theorem 1: (The Right Cancellation Law for Groups)
If $(G, *)$ is a group and $a * b = c * b$ for a, b, and c arbitrary elements of G, then $a = c$.

Premises: $(G, *)$ is a group and $a * b = c * b$ for a, b, and c arbitrary elements of G. Conclusion: $a = c$.

Proof:

Statements	Reasons
1. $a * b = c * b$	1. Premise
2. $b \in G$ means $b' \in G$	2. Axiom G_3 (Inverse $*$)
3. $(a * b) * b' = (c * b) * b'$	3. Definition of $=$
4. $(a * b) * b' = a * (b * b')$ $(c * b) * b' = c * (b * b')$	4. Axiom G_1 (Assoc. $*$)
5. $(a * b) * b' = a * e$ $(c * b) * b' = c * e$	5. Axiom G_3 (Inverse $*$)
6. $(a * b) * b' = a$ $(c * b) * b' = c$	6. Axiom G_2 (Identity $*$)
7. $a = c$	7. Lines 3 and 6 and definition of $=$

As you can see, this process is similar to that of proving theorems in symbolic logic. Our previous experience with the deductive process will be a guide in our present setting. Now we will consider several other theorems in this deductive mathematical system.

Theorem 2: (The Left Cancellation Law for Groups)
If $(G, *)$ is a group and $b * a = b * c$ for any a, b, and c arbitrary elements of G, then $a = c$.

Proof: You try this one (Hint: It is similar to the proof for Theorem 1.)

Does a group $(G, *)$ have more than one identity element relative to $*$? Theorem 3 provides the answer to this question.

Theorem 3: If $(G, *)$ is a group, then there is exactly one identity element e relative to $*$.

Discussion: This theorem is of the form $p \rightarrow q$. To prove this particular theorem, we use a "trick" that is very handy for uniqueness theorems in general; that is, theorems stating that there is exactly one of something. We assume there are at least two distinct identities relative to $*$ and show that this assumption leads to a contradiction. That is, we assume the theorem to be false and show that this leads to a contradiction (a statement of the form $r \wedge \bar{r}$), hence, the theorem must be true. Recall that $(\overline{p \rightarrow q}) \Leftrightarrow (p \wedge \bar{q})$. Hence, to prove the above theorem we

Statements	Reasons
1. e_1 and e_2 are distinct identities relative to $*$; that is, $e_1 \neq e_2$.	1. Premise
2. For $a \in G$, $a * e_1 = a$ and $a * e_2 = a$	2. Axiom G_2 (Identity $*$)
3. $a * e_1 = a * e_2$	3. Definition of $=$
4. $e_1 = e_2$	4. Theorem 2
5. $e_1 = e_2$ and $e_1 \neq e_2$	5. Principle of logic (conjunction of Steps 1 and 4)
6. Contradiction	6. Principle of logic (statement of form $r \wedge \bar{r}$ on Step 5)

have as premises that p: (G,$*$) is a group, and $\bar{q}$: there are at least two distinct identities, e_1 and e_2, relative to $*$. On the basis of these premises we will be able to reach a contradiction.
Hence, we conclude that $p \rightarrow q$ is true. That is, the identity is unique.

The theorem regarding the uniqueness of the inverse of a group element is proved in a similar manner.

Theorem 4: If (G,$*$) is a group and $a \in G$, then there is exactly one inverse element a' for a relative to $*$ such that $a * a' = a' * a = e$.
Proof: Similar to previous theorem.

Theorem 5: Let (G,$*$) be a group and $a \in G$, then $(a')' = a$.
Proof:

Statements	Reasons
1. $a \in G$ means $a' \in G$ and $a' * a = e$.	1. Axiom G_3 (Inverse $*$)
2. $a' \in G$ means $(a')' \in G$ and $a' * (a')' = e$.	2. Axiom G_3 (Inverse $*$)
3. $a' * a = a' * (a')'$	3. Lines 1 and 2 and definition of =
4. $a = (a')'$	4. Theorem 2 (Left Cancellation Law)

Theorem 6: If (G,$*$) is a group and a and b are elements of G, then $(a * b)' = b' * a'$.
Proof:

Statements	Reasons
1. $a,b \in G$ means $a * b \in G$	1. Definition of group ($*$ is a binary operation on G)
2. $a * b \in G$ means $(a * b)' \in G$ and $(a * b) * (a * b)' = e$	2. Axiom G_3 (Inverse $*$)
3. $a,b \in G$ means $a',b' \in G$ and $a * a' = e$ and $b * b' = e$.	3. Axiom G_3 (Inverse $*$)

continued

Statements	Reasons
4. $a', b' \in G$ means $b' * a' \in G$.	4. Definition of group
5. $a * b$ and $b' * a' \in G$ means $(a * b) * (b' * a') \in G$.	5. Definition of group
6. $(a * b) * (b' * a') = a * (b * b') * a'$	6. Axiom G_1 (Assoc. $*$)
7. $(a * b) * (b' * a') = (a * e) * a'$	7. Axiom G_3 (Inverse $*$)
8. $(a * b) * (b' * a') = a * a'$	8. Axiom G_2 (Identity $*$)
9. $(a * b) * (b' * a') = e$	9. Axiom G_3 (Inverse $*$)
10. $(a * b) * (a * b)' = e$ and $(a * b) * (b' * a') = e$	10. Principle of logic (conjunction of lines 2 and 9)
11. $(a * b)' = b' * a'$	11. Theorem 4 (Uniqueness of inverse)

Theorem 7: If $(G, *)$ is a group and e is the identity of G, then $e' = e$.

Proof:

Statements	Reasons
1. $e \in G$ means $e' \in G$ and $e * e' = e$.	1. Axiom G_3 (Inverse $*$)
2. $e * e = e$	2. Axiom G_2 (Identity $*$)
3. $e * e' = e * e$	3. Lines 1 and 2 and definition of $=$
4. $e' = e$	4. Theorem 2 (Left cancellation law)

To help conceptualize the power and economy of work in a deductive system, stop to consider the fact that every group has the properties we have proved so far. Just think of how much work it would have been to show these properties for each individual group representation we have considered! You can now appreciate the respect that a mathematician has for the deductive method.

Once a person becomes comfortable with proofs it is possible not to be so formal. We may adopt a style of writing a proof that tends to show the highlights of the steps in the proof and to omit many of the details, thus making the proof both easier to write and easier to read. However, use of a less formal proof is not a license for careless reasoning. We must always remain alert to any error in our reasoning and not let an error be camouflaged by the careless writing of proofs.

The following theorem will be proved, using a more convenient format.

Theorem 8: If $(G, *)$ is a group and a and b are elements of G, then the equation $a * x = b$ has a unique solution in G.

Proof: The use of a' with the binary operation $*$ on the left of both sides of $a * x = b$ yields $x = a' * b$, which is a solution, since $a * x = a * (a' * b) = (a * a') * b = e * b = b$. To show that there is not more than one solution of $a * x = b$, we assume that there exist x_1 and x_2 such that both are solutions of $a * x = b$. That is, we assume, $a * x_1 = b$ and $a * x_2 = b$. This means $a * x_1 = a * x_2$. But then, by the Left Cancellation Law, $x_1 = x_2$ so that there is exactly one solution of $a * x = b$.

As you can see, the more informal proof style leaves out many of the justifications and does not mention principles of logic involved. That is, only the highlights of the proof are shown. In the exercises that follow, the authors suggest that you do the proofs in the formal format. This will give you a feeling for proofs in this system, since the formal format forces you to be very careful in using the deductive process. After you have become comfortable with using the deductive process in this system, you may use the less formal style.

Exercises 5.2

1. Prove Theorem 2.
2. Prove Theorem 4.
3. Prove that if a and b are elements of a group $(G, *)$,

then there exists a unique element y of $(G, *)$ such that $y * a = b$. Notice that, since for a group $(G, *)$, the equation $a * x = b$ and the equation $y * a = b$ have unique solutions, we can justify the statement made on page 254 that each element of a finite group $(G, *)$ must appear exactly once in each row and column of a group table.

4. Consider the set Z_6^* under the operation of multiplication.

a) Are there two elements a and b from Z_6^* for which the solution of the equation $a \odot x = b$ is not in Z_6^*?

b) Can you find two elements a and b from Z_6^* for which the equation $a \odot x = b$ has two different solutions?

c) Is $(Z_6^*, \odot)$ a group?

5. Prove the following. Let $(G, *)$ be a group. If a and b are elements of $(G, *)$, $[(a * b) * c']' * (a * b) = c$.

6. Notice that Theorem 6 states that in a group $(G, *)$, $(a * b)' = b' * a'$ for all a, b elements of $(G, *)$.

a) Is it the case that $(a \oplus b)' = a' \oplus b'$ for all a, b elements of $(Z_3, \oplus)$?

b) Is it true that for all a and b elements of $(Z, +)$, $(a + b)' = a' + b'$?

c) Find two elements a and b of $(T, \circ)$ where $T = \{f_1, f_2, \ldots, f_6\}$, such that $(a \circ b)' \neq a' \circ b'$. Show that for these same two elements, $(a \circ b)' = b' \circ a'$.

7. Let $(G, *)$ be a group. Prove that if a and b are elements of $(G, *)$ and $a = b$, then $a' = b'$.

8. Let $(G, *)$ be a group. Prove that if a, b, and c are elements of $(G, *)$, then the equation $a * x * b = c$ has a unique solution.

9. Let $(G, *)$ be a group, and let a and b be elements of $(G, *)$ with e the identity of G. Prove that $a * b' = e$, if and only if $a = b$. (We remind the student that to prove this theorem there are two jobs to do: [a] Prove that if $a * b' = e$, then $a = b$, and [b] Prove that if $a = b$, then $a * b' = e$).

10. Let $(G, *)$ be a group. Prove that if a is an element of $(G, *)$ and $a * a = a$, then $a = e$ where e is the identity of $(G, *)$.

Section 3
Subgroups

Recall the group of rigid motions of a rectangle, $(S, \diamondsuit)$, discussed in Chapter 4. We again show the table for this group in Figure 5.7.

The shaded area of the table shows that the subset $K = \{I, R\}$ of S forms a group with respect to $\diamondsuit$. That is, $(K, \diamondsuit)$ is also a group. We further note that, for subsets $K' = \{I, H\}$, $K'' = \{I, V\}$, and $K''' = \{I\}$, we obtain the groups $(K', \diamondsuit)$, $(K'', \diamondsuit)$, and $(K''', \diamondsuit)$, as indicated by the tables in Figure 5.8.

Groups contained within larger groups are of considerable importance in mathematics. We will want to prove some abstract properties about these groups.

Definition: Let H be a nonempty subset of G, and let $(G, *)$ be a group. Then $(H, *)$ is a *subgroup* of $(G, *)$ if and only if H is a group with respect to the binary operation of $(G, *)$.

Example 13: Think of the group $(Z_4, \oplus)$. We see from the table in Figure 5.9 that $(\{0, 2\}, \oplus)$ is a group with respect to $\oplus$.

Hence, $(\{0, 2\}, \oplus)$ is a subgroup of $(Z_4, \oplus)$. Can you find any other subgroups of $(Z_4, \oplus)$?

Example 14: Consider the set $2Z = \{\ldots, -4, -2, 0, 2, 4, \ldots\} \subset Z$. $(2Z, +)$ is a subgroup of $(Z, +)$.

Example 15: Consider the set $\{-1, 1\} \subset Z$. Under multiplica-

◇	I	R	H	V
I	I	R	H	V
R	R	I	V	H
H	H	V	I	R
V	V	H	R	I

Figure 5.7

◇	I	H
I	I	H
H	H	I

Group Table
for (K′,◇)

◇	I	V
I	I	V
V	V	I

Group Table
for (K″,◇)

◇	I
I	I

Group Table
for (K‴,◇)

Figure 5.8

$\oplus$	0	2
0	0	2
2	2	0

Figure 5.9

·	−1	1
−1	1	−1
1	−1	1

Figure 5.10

tion, ({−1,1},·) is a group. The table for this group is shown in Figure 5.10. Note that although ({−1,1},·) is a group, it is not a subgroup of Z under multiplication, since Z under multiplication is not a group.

It is clear that, from the definition of a subgroup, every group has at least two subgroups—the group itself and the group consisting of the identity alone. These two groups are called *improper subgroups*. All other subgroups are called *proper subgroups*.

Theorem 9: If (H,∗) is a subgroup of (G,∗), then the identity of (H,∗) is the identity of (G,∗).

Proof: Since e is the identity of $(G, *)$, then $e * a = a * e = a$ for all $a \in G$. Let i be the identity of $(H, *)$. Since $H \subset G$, this means $i \in G$ and $e * i = i$. But, i the identity of $(H, *)$ means $i * i = i$, so that $e * i = i * i$ and, by the Right Cancellation Law, $e = i$.

We will prove the next statement to be true in a manner similar to the proof for Theorem 9.

Theorem 10: If $(H, *)$ is a subgroup of $(G, *)$, then the inverse of an element a in H is its inverse a' in G.

Proof: $a \in H$ means that $a \in G$. So, since $(G, *)$ is a group, there exists $a_1' \in G$ such that $a * a_1' = e$. But, $(H, *)$ is a group, so that there exists an $a_2' \in H$ such that $a * a_2' = e$. Hence, $a * a_1' = a * a_2'$ and, by the Left Cancellation Law, $a_1' = a_2'$.

The next theorem is a fundamental theorem about subgroups. It is a *characterization theorem* for subgroups; that is, a theorem that offers an alternative way of showing that a subset of a group is a subgroup. You will notice that it is stated in the form of a biconditional (if and only if). Since the biconditional is a theorem (logically true), the component statements in the theorem are equivalent. We know from our principles of logic that we must prove $p \Longrightarrow q$ and $q \Longrightarrow p$ in order to show $p \Longleftrightarrow q$.

Theorem 11: Let $(G, *)$ be a group. A subset H of G is a subgroup of $(G, *)$ if and only if H is nonempty and $a * b'$ is an element of H where a and b are elements of H.

Proof: *Part I:* Prove that, if a subset H of G is a subgroup of $(G, *)$, then H is nonempty and $a * b'$ is an element of H where a and b are elements of H.

Proof: $(H, *)$ is a group since $(H, *)$ is a subgroup of $(G, *)$. Hence, $(H, *)$ must contain an identity element. Thus, H is nonempty. Also, since $(H, *)$ is a group, the inverse of each element b in H is in H; that is, b' is in H. $*$ is a binary operation on H since $(H, *)$ is a group. Hence, for any two elements a and b' (not necessarily distinct) in H, $a * b'$ is in H.

Part II: Prove that, if H is a nonempty subset of G and $a * b'$ is an element of H when a and b are elements of H, then $(H, *)$ is a subgroup of $(G, *)$.

Proof: Since H is not empty, there exists an $a \in H$. Now, by our premises, we know that $a * a' \in H$ for $a \in H$. But $a * a' = e$. Hence, $e \in H$. For any $a \in H$, we apply our premise to $e, a \in H$. Thus, $e * a' \in H$, which means that $a' \in H$. Thus, the inverse of every element in H is in H.

Let $a, b \in H$. We know that $b' \in H$ since $b \in H$. We apply our premise to $a, b' \in H$ to get $a * (b')' \in H$. But $(b')' = b$ and, hence, $a * b \in H$. Hence, $*$ is a binary opera- with respect to H.

Associativity follows from the fact, for any $a, b, c \in H$, that $a, b, c \in G$ and $a * (b * c) = (a * b) * c$ in G so that $a * (b * c) = (a * b) * c$ in H. We have now shown that $(H, *)$ meets all the conditions specified by the definition of a group.

It is sometimes easier to show that a subset H of a group $(G, *)$ is a subgroup of $(G, *)$ by using the criteria stated in the above theorem rather than by using the criteria stated in the definition of a subgroup.

A special subset H of a group $(G, *)$ is defined in the next definition. This subset turns out to be very important in group theory.

Definition: The *center* of a group $(G, *)$, denoted C, is $\{x \mid x \in G \text{ and } g * x = x * g \text{ for all } g \in G\}$.

As the definition indicates, the center of a group is the set of all elements of the group that commute with every element of the group.

Theorem 12: If C is the center of a group $(G, *)$, then $(C, *)$ is a subgroup of $(G, *)$.

Proof: Notice that $e * g = g * e$ for all $g \in G$ so that $e \in C$ and C is not empty. Let $a, b \in C$. This means $a * g = g * a$ for all $g \in G$, and $b * g = g * b$ for all $g \in G$. But, since $b * g = g * b$, then:

$$b' * (b * g) = b' * (g * b)$$
$$g = b' * (g * b)$$
$$g * b' = b' * (g * b) * b'$$
$$g * b' = (b' * g) * (b * b')$$
$$g * b' = b' * g$$

This means $b' \in C$.

Let g be any element of G. Then, $(a * b') * g = a * (b' * g) = a * (g * b') = (a * g) * b' = (g * a) * b' = g * (a * b')$ so that $a * b' \in C$. By Theorem 11, then, $(C, *)$ is a subgroup of $(G, *)$.

Think of the set of functions $T = \{f_1, f_2, \ldots, f_6\}$ along with the binary operation of function composition. We have noted that $(T, \circ)$ is a group. The center of $(T, \circ)$ is $C = \{f_1\}$. If $(G, *)$ is an Abelian group, then the center will be the set G itself.

Exercises 5.3

1. List all the proper subgroups of the following groups.
a) $(Z_4, \oplus)$
b) $(S, \diamondsuit)$; that is, rigid motions of a rectangle
c) $(T, \circ)$; that is, $T = \{f_1, f_2, \ldots, f_6\}$
d) $(Z_3, \oplus)$
e) $(Z_7^*, \odot)$

2. What is the center of the group of rigid motions of a square?

3. What is the center of the group of rigid motions of an equilateral triangle?

4. Find all the subgroups for the group of rigid motions of a square. (There are ten of them. Five of them have two elements and three of them contain four elements.) Find two different subgroups, $(G', \diamondsuit)$ and $(G'', \diamondsuit)$, of order four, such that $(G' \cap G'', \diamondsuit)$ is a group.

5. Prove that if $(H, *)$ and $(K, *)$ are subgroups of the group $(G, *)$, then $(H \cap K, *)$ is a subgroup of $(G, *)$.

6. Give an example to show, for a group $(G, *)$ with subgroups $(H, *)$ and $(K, *)$, that $(H \cup K, *)$ is not necessarily a group.

7. Prove Theorem 12 by using the criterion given in the definition of a subgroup. (It was proved in the text by using the characterization theorem.)

8. Prove that if $(G, *)$ is an Abelian group with identity e, then all elements a of $(G, *)$ where $a * a = e$ form a subgroup of $(G, *)$. (That is, show $H = \{a \mid a * a = e$ for $a \in G\}$ under the operation $*$ is a subgroup of $(G, *)$.)

9. Find the subgroups defined by Problem 8 for each of the following groups.

a) $(S, \diamondsuit)$

b) $(Z_5^*, \odot)$

c) $(Z, +)$

d) $(Q - \{0\}, \cdot)$

10. Let $(G, *)$ be a group and a be an element of G. Define $H_a = \{x \mid x * a = a * x \text{ for all } x \in G\}$; that is, the set of all elements of $(G, *)$ that commute with a. Prove that $(H_a, *)$ is a subgroup of $(G, *)$.

Section 4
Group Generators

The group $(Z_{12}, \oplus)$ has several subgroups. Consider a subset B of Z_{12} that contains the element 2. What other elements of Z_{12} must be in B so that $(B, \oplus)$ is a subgroup of $(Z_{12}, \oplus)$? Since B must be closed with respect to $\oplus$, $2 \oplus 2 = 4 \in B$. But then, $2 \oplus 4 = 6$, $6 \oplus 2 = 8$, $8 \oplus 2 = 10$, and $10 \oplus 2 = 0$ must also be elements of B. From this type of reasoning, we know that 2, 4, 6, 8, 10, and 0 must be in the subgroup, plus any other elements that may arise from combinations of these elements under $\oplus$. For $B = \{0,2,4,6,8,10\}$, it is clear from the table in Figure 5.11 that $(B, \oplus)$ is a group and thus a subgroup of $(Z_{12}, \oplus)$.

In a similar manner, we can begin with more than one element and find a subgroup of $(Z_{12}, \oplus)$ that contains these elements. For instance, consider the elements 6 and 9. Notice that in Z_{12} under $\oplus$, $6 \oplus 9 = 3$, $6 \oplus 6 = 0$, and $9 \oplus 9 = 6$. From this we might suspect that $A = \{0,3,6,9\}$ will have enough elements to provide closure for $\oplus$ in A. To be sure that $\oplus$ is closed with respect to $\oplus$, and to see if $(A, \oplus)$ is a group, we build a table. We then check the table to see if $(A, \oplus)$ has the properties of a group. It turns out that $(A, \oplus)$ is a group and, hence, another subgroup of $(Z_{12}, \oplus)$.

Recall the subgroup $(B, \oplus)$ of $(Z_{12}, \oplus)$ shown in Figure 5.11. Notice that each element of the subgroup can be

$\oplus$	0	2	4	6	8	10
0	0	2	4	6	8	10
2	2	4	6	8	10	0
4	4	6	8	10	0	2
6	6	8	10	0	2	4
8	8	10	0	2	4	6
10	10	0	2	4	6	8

Figure 5.11

expressed in terms of the element 2. That is,

$$2 \oplus 2 = 4$$
$$2 \oplus 2 \oplus 2 = 6$$
$$2 \oplus 2 \oplus 2 \oplus 2 = 8, \text{ etc.}$$

Under these conditions we say that 2 is a *generator* of the subgroup $(B, \oplus)$. It is convenient to use the following notational scheme.

$$2^2 = 2 \oplus 2 = 4$$
$$2^3 = 2 \oplus 2 \oplus 2 = 6$$
$$2^4 = 2 \oplus 2 \oplus 2 \oplus 2 = 8, \text{ etc.}$$

We speak of 2^2, 2^3, 2^4, etc., as *positive powers* of the element 2 of $(Z_{12}, \oplus)$. We generalize this to any group $(G, *)$ by noting that for $a \in G$, the nth positive power of a is $a^n = \underbrace{a * a * \ldots * a}_{n \text{ times}}$.

Definition: Let $(G, *)$ be a finite group. $(K, *)$ is the subgroup of $(G, *)$ *generated* by the elements $a_1, a_2, \ldots, a_k$ of G, if and only if every element of the subgroup $(K, *)$ can be expressed in terms of a combination of positive powers of these elements.

Notice, for $2 \in Z_{12}$, that $2^6 = 0 \in Z_{12}$ under $\oplus$ for Z_{12}. Six is the smallest positive power of 2 such that 2 to that positive power is the identity of $(Z_{12}, \oplus)$. We generalize this idea in the following definition.

Definition: For a group $(G,*)$ and an element a of $(G,*)$, if there exists a natural number n such that $a^n = e$, then the least such natural number is the *order of a*. If no such n exists, a is said to be of *infinite order*.

Example 16: Consider the group of rigid motions of a rectangle, $(S,\diamond)$. $(\{I, H\},\diamond)$ is a subgroup of $(S,\diamond)$, and $H^2 = I$, so that the order of H is 2, and H is a generator of $(\{I, H\},\diamond)$.

Example 17: The element 3 will generate $(Z_5^*,\odot)$, since:

$$3^1 = 3$$
$$3^2 = 3 \odot 3 = 4$$
$$3^3 = 3 \odot 3 \odot 3 = 2$$
$$3^4 = 3 \odot 3 \odot 3 \odot 3 = 1$$

The order of 3 is 4 since 1 is the identity of $(Z_5^*,\odot)$, $3^4 = 1$, and there is no positive number less than 4 such that this positive power of 3 is 1.

Example 18: $(Z,+)$ is a group such that 0 is the identity. Notice, for any $z \in Z$ except $z = 0$, that $z^n \neq 0$. Hence, the order of every element of Z except 0 is infinite.

Notice that the positive powers of 2 do not generate a subgroup of $(Z,+)$. That is, 2, $2 + 2 = 4$, $2 + 2 + 2 = 6, \ldots, 2^n, \ldots$ do not form a group with respect to addition. However, if we take 2 and its inverse, -2, then $2 + (-2) = 0$, $-2 + (-2) = -4$, $-2 + (-2) + (-2) = -6$, etc., so that 0 and all the negative numbers of 2Z are determined. Now, $(2Z,+)$ is a subgroup of $(Z,+)$. We generalize this by saying that, when $(G,*)$ is an infinite group, $(K,*)$ is a subgroup of $(G,*)$ generated by $a_1, a_2, \ldots, a_k$ of G, if and only if every element of $(K,*)$ can be expressed in terms of a combination of positive powers of these elements and their inverses.

Let us consider the subgroups of some of the groups we have studied. From our example, we saw that $(\{0, 2\},\oplus)$ was a subgroup of $(Z_4,\oplus)$. Notice that the order of Z_4 is 4 and the order of $(\{0, 2\},\oplus)$ is 2. That is, the order of the subgroup $(\{0, 2\},\oplus)$ divides the order of the group $(Z_4,\oplus)$. Consider the group $(S,\diamond)$. We saw that this group had several subgroups besides the improper ones. $(K, \Diamond)$, $(K', \Diamond)$, and $(K'', \Diamond)$ were all subgroups of order two. Again we note that the order of these subgroups of $(S, \Diamond)$ divides the order of the group $(S, \Diamond)$. Think of $(T, \circ)$. The set $A = \{f_1, f_2, f_3\}$ under function composition is a subgroup of $(T, \circ)$. Let $B = \{f_1, f_4\}$, $C = \{f_1, f_5\}$, and $D = \{f_1, f_6\}$. Then, $(B, \circ)$, $(C, \circ)$, and $(D, \circ)$ are all subgroups of $(T, \circ)$. Again we notice that all the proper subgroups of $(T, \circ)$ have orders that divide the order of $(T, \circ)$. You may wonder if a general statement can be made regarding the order of a group and the order of any of its subgroups. We state *Lagrange's Theorem** in answer to this question. The proof of this theorem involves more machinery than we desire to develop in this text, so it will be omitted. Perhaps, as you study this subject further in some of the materials suggested by our reading list, you will uncover the elegant proof of this theorem.

Theorem 13: **(Lagrange's Theorem)**
Let $(G, *)$ be a group and $(H, *)$ be a subgroup of $(G, *)$. Then, the order of $(H, *)$ divides the order of $(G, *)$.

Exercises 5.4

1. What is the order of the elements given in the following groups?

a) 2 in $(Z_4, \oplus)$
b) 1 in $(Z_4, \oplus)$
c) f_3 in $(T, \circ)$
d) 0 in $(Z_3, \oplus)$
e) 1 in $(Z, +)$
f) 1 in $(Z_5^*, \odot)$
g) R in $(S, \Diamond)$
h) e in $(G, *)$
i) 3 in $(Z, +)$

2. For the groups mentioned in Problem 1, what is the order of the respective subgroup generated by the given element? Can you make a conjecture as to the relationship of the order of an element and the order of the subgroup that it generates?.

*Lagrange's Theorem is named for the French mathematician Joseph Louis Lagrange (1736–1813).

3. Which elements are generators of the following subgroups of $(Z_6, \oplus)$?

a) $(\{0\}, \oplus)$

b) $(Z_6, \oplus)$

c) $(\{0, 2, 4\}, \oplus)$

d) $(\{0, 3\}, \oplus)$

4. Consider $(T, \circ)$. What subgroup does the set $\{f_3, f_4\}$ generate?

5. Find all the generators for each of the following groups.

a) $(Z_5, \oplus)$

b) $(Z_3^*, \odot)$

c) $(Z_{16}, \oplus)$

d) $(Z_{36}, \oplus)$

6. Recall the group of rigid motions of a rectangle, $(S, \diamond)$. Does this group have a generator? Does the subgroup $(\{I, V\}, \diamond)$ have a generator?

7. Consider the set $(\ldots, -6, -3, 0, 3, 6, \ldots\} = 3Z$. Under the operation of addition, what elements generate $3Z$?

8. For the group $(Z_{16}, \oplus)$, what are the possible orders of the subgroups of $(Z_{16}, \oplus)$?

Section 5

Isomorphisms of a Group

Think of the group models $(Z_2, \oplus)$ and $(\{-1, 1\}, \cdot)$. The tables for these groups are shown in Figures 5.12 and 5.13, respectively.

Suppose we construct the following correspondence for elements of these two groups.

$$1 \leftrightarrow 0$$
$$-1 \leftrightarrow 1$$

Now, let us take the table in Figure 5.13 and associate the binary operation of $\cdot$ with $\oplus$, using the above indicated correspondence of the elements. We present in the table in Figure 5.14 the result of this association and correspondence. Notice that the table in Figure 5.14 is either the table in Figure 5.12 or in Figure 5.13, depending on which operation and corresponding set of elements you choose. This means that the groups $(Z_2, \oplus)$ and $(\{-1, 1\}, \cdot)$ are the same, and that one may be regarded as a relabeling of the other. We say groups related in the above manner are *isomorphic.* The word *isomorphic* is derived from the prefix *iso,* meaning "equal," and the root *morph,* meaning "form." That is, the two groups are of the same form or structure. The idea of isomorphism is a very important one in mathematics, so we shall pursue it further, making use of what we have learned about functions to facilitate our development.

⊕	0	1
0	0	1
1	1	0

Figure 5.12

·	1	−1
1	1	−1
−1	−1	1

Figure 5.13

Definition: Let (G,∗) and (G′,#) be groups. A function f that is a one-to-one onto function from G to G′ is called an *isomorphism,* if and only if f(*a* ∗ *b*) = f(*a*) # f(*b*) for every *a*, *b*, ∈ G. In this case, (G,∗) and (G′,#) are said to be *isomorphic,* denoted (G,∗) ≅ (G′,#).

Consider again the groups $(\{-1, 1\}, \cdot)$ and $(Z_2, \oplus)$. In Figure 5.15, we specify a one-to-one onto function from $\{-1, 1\}$ to Z_2. We will check to see if $f(a \cdot b) = f(a) \oplus f(b)$ for all possible choices of a and b in $\{-1, 1\}$.

a) $f(-1 \cdot 1) = f(-1) = 1$
$f(-1) \oplus f(1) = 1 \oplus 0 = 1$
means $f(-1 \cdot 1) = f(-1) \oplus f(1)$

⊕ ↔ ·	0 ↔ 1	1 ↔ −1
0 ↔ 1	0 ↔ 1	1 ↔ −1
1 ↔ −1	1 ↔ −1	0 ↔ 1

Figure 5.14

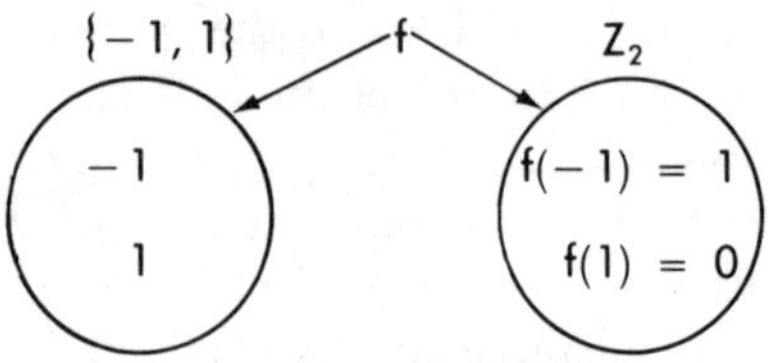

Figure 5.15

b) $f(-1 \cdot -1) = f(1) = 0$
$f(-1) \oplus f(-1) = 1 \oplus 1 = 0$
means $f(-1 \cdot -1) = f(-1) \oplus f(-1)$

c) $f(1 \cdot -1) = f(-1) = 1$
$f(1) \oplus f(-1) = 0 \oplus 1 = 1$
means $f(1 \cdot -1) = f(1) \oplus f(-1)$

d) $f(1 \cdot 1) = f(1) = 0$
$f(1) \oplus f(1) = 0 \oplus 0 = 0$
means $f(1 \cdot 1) = f(1) \oplus f(1)$

Now, since $f(a \cdot b) = f(a) \oplus f(b)$ for all possible choices of $a, b \in \{-1, 1\}$, we see that f is an isomorphism and, hence, that $(\{-1, 1\}, \cdot) \cong (Z_2, \oplus)$. This result, which utilizes our sophisticated machinery, should not surprise us, because we already knew the groups were exactly alike except for the binary operations and the names of the elements of the two groups from Figure 5.14. Had we defined f in such a way that $f(-1) = 0$ and $f(1) = 1$, then f would not have been an isomorphism, since $f(-1 \cdot 1) = f(-1) = 0$ and $f(-1) \oplus f(1) = 0 \oplus 1 = 1$; hence, $f(-1 \cdot 1) \neq f(-1) \oplus f(1)$. It is clear that if two groups, $(G, *)$ and $(G', \#)$, are isomorphic, then G and G′ must be equivalent. But some or all of the one-to-one onto functions between equivalent sets may not be isomorphisms. To show that two groups are isomorphic, all we need to do is find one one-to-one onto function that is an isomorphism between them.

We might suspect that if two groups are isomorphic, their identity elements must correspond and their inverse elements must also correspond.

Theorem 14: Let $(G, *)$ and $(G', \#)$ be isomorphic groups. If e is the identity of $(G, *)$, then f(e) is the identity of $(G', \#)$.

Proof: Since $(G, *)$ and $(G', \#)$ are isomorphic, there exists a one-to-one onto function from G to G′, such that $f(a * b) = f(a) \# f(b)$ for all $a, b \in G$. In particular, consider $a, e \in G$. Then, $a = a * e$, so that $f(a) = f(a * e) = f(a) \# f(e)$. Thus, $f(a) \# f(e) = f(a)$ for $f(a), f(e) \in G'$, and, since the identity of a group is unique, f(e) is the identity of $(G', \#)$.

Theorem 15: Let $(G, *)$ and $(G', \#)$ be isomorphic groups. If

a' is the inverse of $a \in G$, then $f(a')$ is the inverse of $f(a)$ in $(G', \#)$.

Proof: Let f be an isomorphism between $(G, *)$ and $(G', \#)$. Then, $f(e) = f(a * a') = f(a) \# f(a')$, so that $f(a')$ is the inverse of $f(a)$ in $(G', \#)$.

Another important property of an isomorphism between two groups $(G, *)$ and $(G', \#)$ is that if $a_1 \in G$ and $a_2 \in G'$ such that a_1 corresponds to a_2 under the isomorphism, then the nth positive power of a_1 corresponds to the nth positive power of a_2. If we are trying to associate the elements of two groups in such a manner as to display an isomorphism between them, it is helpful to keep this fact in mind.

Theorem 16: Let $(G, *)$ and $(G', \#)$ be isomorphic groups such that $a_1 \in G$ corresponds with $a_2 \in G'$ under the isomorphism. Then, the nth positive power of a_1 corresponds to the nth positive power of a_2.

Discussion: Let f be the isomorphism between $(G, *)$ and $(G', \#)$. Since $a_1 \in G$ corresponds with $a_2 \in G'$, we know that $f(a_1) = a_2$. Hence, we note the following pattern.

a_1 under f is $f(a_1) = a_2$

$$a_1^2 = a_1 * a_1, \text{ so that } f(a_1 * a_1) = f(a_1) \# f(a_1) = a_2 \# a_2 = a_2^2$$

$$a_1^3 = a_1 * a_1 * a_1, \text{ so that } f(a_1 * a_1 * a_1) = f(a_1) \# f(a_1) \# f(a_1) = a_2 \# a_2 \# a_2 = a_2^3$$

$$\vdots$$

$$a_1^n = \overbrace{a_1 * a_1 * \dots * a_1}^{n \text{ times}}, \text{ so that } f(\overbrace{a_1 * a_1 * \dots * a_1}^{n \text{ times}})$$

$$= \underbrace{f(a_1) \# f(a_1) \# \dots \# f(a_1)}_{n \text{ times}} = \underbrace{a_2 \# a_2 \# \dots \# a_2}_{n \text{ times}} = a_2^n$$

It is clear from the preceding discussion that the nth positive power of a_1 corresponds with the nth positive power of a_2.

Example 19: Consider the groups $(Z_4, \oplus)$ and $(Z_5^*, \odot)$. From the previous three theorems, it is clear that if these two groups

Elements of $(Z_4, \oplus)$		Elements of $(Z_5^*, \odot)$
0	⟷	1
1	⟷	3
2	⟷	4
3	⟷	2

Figure 5.16

are isomorphic, then $0 \in Z_4$ must correspond to $1 \in Z_5^*$, since these are the identities of their respective groups. Since the *n*th positive powers of elements of isomorphic groups must correspond, corresponding elements are of the same order. We note that:

Elements of $(Z_4, \oplus)$	*Elements of* $(Z_5^*, \odot)$
$1^4 = 1 \oplus 1 \oplus 1 \oplus 1 = 0$	$2^4 = 2 \odot 2 \odot 2 \odot 2 = 1$
$2^2 = 2 \oplus 2 = 0$	$3^4 = 3 \odot 3 \odot 3 \odot 3 = 1$
$3^4 = 3 \oplus 3 \oplus 3 \oplus 3 = 0$	$4^2 = 4 \odot 4 = 1$

Hence, we see Figure 5.16 pictures a correspondence between the elements of $(Z_4, \oplus)$ and $(Z_5^*, \odot)$, which displays an isomorphism. We notice that under the correspondence given in Figure 5.16, inverse corresponds to inverse. For instance, 3 is the inverse of 1 in $(Z_4, \oplus)$, and the associated elements 2 and 3 in $(Z_5^*, \odot)$ are such that 2 is the inverse of 3. By checking the group tables for $(Z_4, \oplus)$ and $(Z_5^*, \odot)$ and making the above indicated associations, we notice that the tables are identical and, hence, $(Z_4, \oplus)$ and $(Z_5^*, \odot)$ are isomorphic. The isomorphism displayed in Figure 5.16 is not the only isomorphism that shows $(Z_4, \oplus)$ and $(Z_5^*, \odot)$ to be isomorphic groups. Can you find another one?

Example 20: Is the group of rigid motions of a rectangle $(S, \diamond)$, shown in Figure 5.17, isomorphic to the group displayed in Figure 5.18? It is clear that I is the identity of the group table in Figure 5.17 and *x* is the identity of the group displayed in Figure 5.18. Hence, if these groups are isomorphic, I must correspond

◇	I	R	H	V
I	I	R	H	V
R	R	I	V	H
H	H	V	I	R
V	V	H	R	I

Figure 5.17

with x. We also note that:

$$R^2 = I \qquad w^2 = x$$
$$H^2 = I \qquad y^2 = x$$
$$V^2 = I \qquad z^2 = x$$

The following correspondence displays one isomorphism between these two groups.

$$I \leftrightarrow x$$
$$R \leftrightarrow w$$
$$H \leftrightarrow y$$
$$V \leftrightarrow z$$

There are other isomorphisms that can be specified. See if you can determine some of them.

Example 21: $(Z_3^*, \odot)$ and $(\{-1, 1\}, \cdot)$ are isomorphic. This is verified by the following isomorphism.

Elements of $(Z_3^*, \odot)$		Elements of $(\{-1, 1\}, \cdot)$
1	$\longleftrightarrow$	1
2	$\longleftrightarrow$	-1

*	x	w	y	z
x	x	w	y	z
w	w	x	z	y
y	y	z	x	w
z	z	y	w	x

Figure 5.18

Example 22: The groups $(Z,+)$ and $(2Z,+)$ can be shown to be isomorphic by specifying a one-to-one onto function $f:Z \rightarrow 2Z$ such that $f(a + b) = f(a) + f(b)$ for all $a, b \in Z$. We specify f as follows: For $z \in Z$, then $f(z) = 2z$. To make sure that f is an isomorphism, we must:

a) be sure that f is a function. That is, show that if $z_1, z_2 \in Z$ and $z_1 = z_2$, then $f(z_1) = f(z_2)$.

b) be sure that f is one-to-one. That is, show that if $f(z_1)$ and $f(z_2)$ are elements of 2Z such that $f(z_1) = f(z_2)$, then $z_1 \doteq z_2$.

c) be sure that f takes Z onto 2Z. That is, show that for any $x \in 2Z$, that $\exists\, z \in Z$ such that $f(z) = x$.

d) be sure that for any $z_1, z_2 \in Z$, $f(z_1 + z_2) = f(z_1) + f(z_2)$.

We verify the above four properties for f as follows.

a) If $z_1 = z_2$, then $2z_1 = 2z_2$. But $f(z_1) = 2z_1$ and $f(z_2) = 2z_2$, so that $f(z_1) = f(z_2)$ and f is a function.

b) If $f(z_1) = f(z_2)$, then $2z_1 = 2z_2$. But, dividing by 2, we have $z_1 = z_2$, and f is one-to-one.

c) Suppose x is any element of 2Z. Then, since every element of 2Z is even, x can be written in the form $2y$ for $y \in Z$. Hence, for any element x in 2Z, there exists an element y in Z such that $x = 2y$; that is, $y = \frac{x}{2}$, and f takes Z onto 2Z.

d) $f(z_1 + z_2) = 2(z_1 + z_2) = 2z_1 + 2z_2 = f(z_1) + f(z_2)$, and f preserves the operation between Z and 2Z.

Hence, $(Z,+)$ and $(2Z,+)$ are isomorphic.

When we can directly display the isomorphism between groups, as in Examples 19, 20 and 21, we do not resort to such a fancy discussion as in Example 22. In Example 22, it is clear how the function concept is central to the discussion of isomorphism.

Example 23: We will now show two groups that are of the same order but which are not isomorphic. Consider the groups $(T,\circ)$ and $(Z_6, \oplus)$. Recall that $T = \{f_1, f_2, \ldots, f_6\}$ and $\circ$ is the operation of function composition. One cannot find an isomorphism between these two groups. We should expect this to be the case, since the two groups are structually different. That is, $(Z_6, \oplus)$ is abelian, but $(T,\circ)$ is not.

Exercises 5.5

1. Specify a one-to-one pairing of elements that would indicate that the following groups are isomorphic.

a) $(Z_3^*, \odot)$ and $(Z_2, \oplus)$

b) $(T, \circ)$ and the group of rigid motions of an equilateral triangle.

c) $(Z_4, \oplus)$ and the group of rotations of a square.

d) $(M, \circ)$ and the group of rigid motions of a square where $(M, \circ)$ is given as follows: Consider the set of permutations on a set of four elements. A subgroup of this group of permutations is specified as follows.

$$A = \begin{pmatrix} 1 & 2 & 3 & 4 \\ 1 & 2 & 3 & 4 \end{pmatrix} \qquad B = \begin{pmatrix} 1 & 2 & 3 & 4 \\ 3 & 4 & 1 & 2 \end{pmatrix}$$

$$C = \begin{pmatrix} 1 & 2 & 3 & 4 \\ 4 & 3 & 2 & 1 \end{pmatrix} \qquad D = \begin{pmatrix} 1 & 2 & 3 & 4 \\ 2 & 1 & 4 & 3 \end{pmatrix}$$

$$E = \begin{pmatrix} 1 & 2 & 3 & 4 \\ 3 & 2 & 1 & 4 \end{pmatrix} \qquad F = \begin{pmatrix} 1 & 2 & 3 & 4 \\ 1 & 4 & 3 & 2 \end{pmatrix}$$

$$G = \begin{pmatrix} 1 & 2 & 3 & 4 \\ 2 & 3 & 4 & 1 \end{pmatrix} \qquad H = \begin{pmatrix} 1 & 2 & 3 & 4 \\ 4 & 1 & 2 & 3 \end{pmatrix}$$

Let $M = \{A, B, C, D, E, F, G, H\}$, and let the binary operation on M be function composition. Then, $(M, \circ)$ is a group.

2. Consider the two groups of order four, $(Z_4, \oplus)$ and $(S, \diamond)$. Are these two groups isomorphic?

3. Show that $(Z, +)$ and $(3Z, +)$ are isomorphic.

4. Prove that groups $(G, *)$ and $(G', *)$ are isomorphic only if $n(G) = n(G')$.

5. If $(G, *)$ is an infinite group and $(G', *)$ is a finite group, then $(G, *)$ and $(G', *)$ cannot be isomorphic. Why?

6. Prove, if $(G, *)$ is finite and $(H, *)$ is a subgroup of $(G, *)$ such that H is a proper subset of G, that $(G, *)$ and $(H, *)$ are not isomorphic.

7. a) Find a group $(G, *)$ isomorphic to $(Z_4, \oplus)$ different from $(Z_4, \oplus)$.

b) Find a group $(G', \#)$ isomorphic to $(S, \diamondsuit)$ different from $(S, \diamondsuit)$.

c) Are the groups $(G, *)$ and $(G', \#)$ isomorphic?

8. Let f be an isomorphism of a group $(G, *)$ with a group $(G', \#)$. If $(H, *)$ is a subgroup of $(G, *)$ and $H' = \{f(x) \mid x \in H\}$, show that $(H', \#)$ is a subgroup of $(G', \#)$.

9. Prove, if $(G, *)$ is an Abelian group, that the function f from G to G specified by $f(a) = a'$ for $a \in G$ is an isomorphism.

Section 6
More Basic Theorems about Groups

We hope you are enjoying the power and beauty of this abstract deductive mathematical system. We will prove a few more theorems from the vast literature in this area and then provide you with an opportunity to enjoy doing some more theorems for yourself in the exercises.

Theorem 17: Let $(G, *)$ be a group, and let a be a fixed element of $(G, *)$. Let $f: G \rightarrow G$ be specified by the following rule of correspondence.

$$f(x) = a * x * a' \text{ for } x \in G$$

Then, f is an isomorphism of G with itself.

Proof: f is a function, since if $x_1 = x_2$, then $f(x_1) = a * x_1 * a' = a * x_2 * a' = f(x_2)$. f is one-to-one, for if $f(x_1) = f(x_2)$, then $a * x_1 * a' = a * x_2 * a'$ and by the Left and Right Cancellation Laws for groups, $x_1 = x_2$. For any $y \in G$ we have $a' * (a * y * a') * a = y$, so that f is from G onto G. Let $x, y \in G$. Then, $f(x * y) = a * (x * y) * a' = a * (x * e * y) * a' = a * (x * a' * a * y) * a' = (a * x * a') * (a * y * a') = f(x) * f(y)$. Hence, f is an isomorphism of G with itself.

We are familiar with the property of equality, which says that if $a = b$ and $b = c$, then $a = c$. A similar property holds for isomorphic groups.

Theorem 18: If (G,$*$) and (G$'$,#) are isomorphic and (G$'$,#) and (G$''$,¢) are isomorphic, then (G,$*$) and (G$''$,¢) are isomorphic.

Proof: Since (G,$*$) and (G$'$,#) are isomorphic, this means there exists an isomorphism f between G and G$'$: that is, f:G $\rightarrow$ G$'$ is a one-to-one correspondence from G to G$'$ such that for a, $b \in$ G, f($a * b$) = f(a) # f(b). Similarly, there exists an isomorphism g between G$'$ and G$''$: that is, g:G$'$ $\rightarrow$ G$''$ is a one-to-one correspondence from G$'$ to G$''$such that for x, $y \in$ G$'$, g(x # y) = g(x) ¢ g(y). Since both f and g are one-to-one correspondences, we know that g $\circ$ f is a one-to-one correspondence from G to G$''$ (Exercise 4.8, Problem 6,7). For a, $b \in$ G, we have g $\circ$ f ($a * b$) = g(f($a * b$)) = g(f(a) # f(b)) = g(f(a)) ¢ g(f(b)) = g $\circ$ f (a) ¢ g $\circ$ f (b). Hence g $\circ$ f is an isomorphism from (G,$*$) to (G$''$,¢) so that (G,$*$) is isomorphic to (G$''$,¢).

If one of two isomorphic groups is known to be commutative, the following theorem assures us that the other group is also commutative. This is only reasonable, since isomorphic groups are structurally the same. A structural property of a group is a property that can be expressed in terms of the binary operation of a group and is not dependent on the specific nature of the objects of the group. Hence, it is clear that a structural property of a group is reflected by the same structural property "relabeled" in any isomorphic group.

Theorem 19: Let (G,$*$) and (G$'$,#) be isomorphic groups such that (G,$*$) is abelian. Then (G$'$,#) is abelian.

Proof: Let f be the isomorphism between G and G$'$. Since (G,$*$) is commutative, then ($a * b$) = ($b * a$) for a, $b \in$ G, and, hence, f($a * b$) = f($b * a$). Then for f(a), f(b) $\in$ G$'$, we have f(a) # f(b) = f($a * b$) = f($b * a$) = f(b) # f(a) so that (G$'$,#) is commutative.

Exercises 5.6

1. We know that an element b of a group (G,$*$) is the identity of (G,$*$) if for all a element of (G,$*$), $a * b = a$. Prove,

if (G, $*$) is a group and $a * b = a$ for some a element of (G, $*$), that b is the identity of (G, $*$).

2. Let (G, $*$) be a group and H a nonempty subset of G. Prove, if $*$ is a binary operation on H and a' is an element of H for all $a \in$ H, that (H, $*$) is a subgroup of (G, $*$).

3. Prove, if (G, $*$) is a group such that for all a elements of (G, $*$), $a * a = e$, that (G, $*$) is abelian.

4. Show, if (G, $*$) is a group of 2, 3, or 4 elements, that (G, $*$) is Abelian.

5. Prove, if (G, $*$) is a group such that for all a and b elements of (G, $*$), $(a * b) * (a * b) = (a * a) * (b * b)$, that (G, $*$) is abelian.

6. Find four subgroups of the positive rational numbers under multiplication.

7. Prove, if the function f from G to G specified by $f(a) = a'$ for $a \in$ G is an isomorphism, that (G, $*$) is an abelian group.

Section 7
Concluding Remarks

As we look back over the material in this chapter, as well as that in the preceding chapters, we see many ideas fitting together to produce a powerful and beautiful system. It reminds us of seeing a master artist paint a picture. At first the brush strokes and various colors on the canvas may seem to have no apparent design. Then, as the work progresses, we become aware of an emerging pattern and realize that all of the varied colored strokes of the artist's brush relate to produce a masterpiece. Working with the particular deductive mathematical system called a group, we have brought to bear on our endeavors most of the previous concepts presented in this book. One cannot help but appreciate the beauty and enjoyment inherent in such mathematical activities.

We have relied heavily on the deductive process in our work. The formal approach to proving theorems has been replaced with a warmer, more easily flowing manner of stating proofs, which gives mainly the highlights of a sequence of statements that would appear in a formal proof. This facilitates both the writing and the reading of proofs for theorems.

Among the different models or representations of a group, some were familiar, for instance, $(Z, +)$, $(Q, +)$, $(\mathbf{R}, +)$, $(Q - \{0\}, \cdot)$, and $(\mathbf{R} - \{0\}, \cdot)$. Other models were less familiar or, perhaps, completely new. Yet, we gathered all of these different systems into one camp and discussed them in terms of the abstract notion of a group. This is truly magnificent. As the artist enjoys creating a masterpiece, so the mathematician enjoys a similar personal aesthetic satisfaction. Not only does the artist enjoy his masterpiece, but others enjoy and appreciate it also. Similarly, the mathematician does not want to keep his work hidden in a corner, but rather desires that others should be able to appreciate it. Learning to appreciate the intellectual activities of a scholar in any discipline is part of an education. All humanity benefits from the mathematician's development of a general theory that unifies various areas of endeavor.

References
Chapter Five

Dinkines, Flora, *Abstract Mathematical Systems* (New York: Appleton-Century-Crofts, 1964), pp. 24–28.

Fraleigh, John B., *A First Course in Abstract Algebra* (Menlo Park, Calif.: Addison-Wesley Publishing Company, 1968), pp. 80–86.

Phillips, John L., *The Origins of Intellect* (San Francisco: W. H. Freeman and Company, 1969), pp. 67–75.

Weiner, Louis M., *Introduction to Modern Algebra* (Atlanta: Harcourt, Brace and World, Inc., 1970), pp. 91–122.

Newman, James R., ed., *The World of Mathematics*, Vol. 2 (New York: Simon and Schuster, 1956), pp. 1294–1313.

Zwier, Paul J., and Nyhoff, Larry R., *Essentials of College Mathematics* (San Francisco: Holt, Rinehart and Winston, Inc., 1971), pp. 212–218.

CHAPTER SIX

Rings and Fields

We considered the abstract mathematical system called a group in the previous chapter. In the study of groups, we considered only one binary operation on a given set. In this chapter, we consider abstract systems that have one binary operation on a set with the properties of an abelian group and another binary operation on the same set with at least some (and possibly all) of those properties. The two operations are related by a distributive property, and the particular abstract systems are deductive mathematical systems called *rings, commutative rings with identity, integral domains,* and *fields*. To progress from rings to fields, we add axioms about the second binary operation.

We will build these new deductive mathematical systems upon the foundation laid in the study of groups by using new definitions, such as the definition of a ring. We will use the deductive method to prove theorems in these new systems. Various representations of rings are given, including some from the familiar number systems. Again, the concept of sets provides continuity in our study of various mathematical systems.

In the previous chapter on groups, the symbol $*$ was used to denote the binary operation on a set. We dealt not only with the binary operation but also with the elements in a set in abstract form. Similarly, in this chapter we will deal with the binary operations and the elements of a set in an abstract manner.

Section 1
Rings

You have been initiated into thinking of binary operations in an abstract manner, and with the mathematical maturity that you now possess, we can adopt some conventional simplifications of notation. We will call the two ring operations addition (our first binary operation) and multiplication (our second binary operation), even though the ring operations may not be addition and multiplication.

Definition: A nonempty set R, together with two binary operations, addition, $+$, and multiplication, $\cdot$, defined on R is called a *ring*, denoted $(R, +, \cdot)$, if and only if:

a) $(R, +)$ is an abelian group;

b) multiplication is associative in R;

c) for all $a, b, c \in R$, $a \cdot (b + c) = (a \cdot b) + (a \cdot c)$—the *left distributive law*—and $(b + c) \cdot a = (b \cdot a) + (c \cdot a)$—the *right distributive law*.

In the remainder of this text, we will adopt some notational devices commonly used in the number systems. However, for the sake of clarity, there will be instances in which we do not follow these conventions.

a) The symbol 0 (zero) will denote the identity for addition.

b) $a + b \cdot c$ will mean $a + (b \cdot c)$.

c) The additive inverse of a will be denoted $-a$.

d) $a \cdot b$ will be denoted by ab.

e) $a + (-b)$ will be written $a - b$.

Notice that in additive notation, $(a')' = -(-a)$, and $(a + b)' = -(a + b)$. Therefore, in additive notation, the group properties of the inverse stated in Theorems 5 and 6 of the previous chapter (page 262) are $-(-a) = a$ and $-(a + b) = -b + (-a) = -b - a$.

Definition: A ring $(R, +, \cdot)$ is *commutative* if and only if $ab = ba$ for all $a, b \in R$.

Definition: A ring $(R, +, \cdot)$ has an *identity* element if and only if there exists an element e of R such that $ea = ae = a$ for all $a \in R$.

In accordance with our adoption of notation from the number systems, we will denote the multiplicative identity of a ring by 1. When we refer specifically to a ring with identity we are referring to a ring with a multiplicative identity, since every ring has an additive identity. If an element a of a ring has a multiplicative inverse, we will denote this inverse by a^{-1}.

Example 1: The familiar number systems $(\mathbf{R}, +, \cdot)$, $(Q, +, \cdot)$, and $(Z, +, \cdot)$ are all rings.

Example 2: Let Z_m, $\oplus$, and $\odot$ be defined as in the previous chapter (page 234). Then $(Z_m, \oplus, \odot)$ is a ring. Notice that $(Z_m, \oplus, \odot)$ can be written $(Z_m, +, \cdot)$ under our adopted convention.

Example 3: Let $A = \{a, b\}$, with addition and multiplication defined by the tables in Figure 6.1. $(A, +, \cdot)$ is a ring.

Notice that the rings of Examples 1, 2, and 3 are all commutative rings with an identity element.

+	a	b
a	a	b
b	b	a

·	a	b
a	a	a
b	a	b

Figure 6.1

Example 4: Let A = {*a*, *b*, *c*, *d*}, with addition and multiplication specified by the tables in Figure 6.2. (A, +, ·) is a ring that is not commutative since bc = c but cb = a. Notice that (A, +, ·) does not have an identity.

+	a	b	c	d
a	a	b	c	d
b	b	a	d	c
c	c	d	a	b
d	d	c	b	a

·	a	b	c	d
a	a	a	a	a
b	a	b	a	b
c	a	c	a	c
d	a	d	a	d

Figure 6.2

Example 5: Consider (2Z, +, ·). This is a commutative ring, but it has no identity element.

Example 6: Let (G, +) be any Abelian group. Define ·, a binary operation on G, by $ab = 0$ for all $a, b \in$ G. This is a commutative ring that does not have an identity element.

The distributive laws are the only means we have of relating the binary operations of a ring. Hence, it is important to realize that both the left and right distributive laws must hold in a system if it is a ring. In the following examples, we see structures that have all of the properties of a ring except that one of the distributive laws does not hold. (Examples 7 and 8 are taken from the article "Is X·0 = 0 Always True?" by R. Hunkler and H. Heatherly in the November 1971 issue of *The Mathematics Teacher.*)

Example 7: Let A = {0, 1, 2} and define + and · as in the tables in Figure 6.3. Notice that $(1 + 2)\cdot 1 = 0\cdot 1 = 1$ and $1\cdot 1 + 2\cdot 1 = 1 + 1 = 2$ so that $(1 + 2)\cdot 1 \neq 1\cdot 1 + 2\cdot 1$. Hence, the right distributive law fails although the left distributive law holds.

Example 8: Let A = {0, 1, 2, 3} and + and · be defined according to the tables in Figure 6.4

In this example the left distributive law fails while the right distributive law holds.

+	0	1	2
0	0	1	2
1	1	2	0
2	2	0	1

·	0	1	2
0	0	0	0
1	1	1	1
2	2	2	2

Figure 6.3

+	0	1	2	3
0	0	1	2	3
1	1	2	3	0
2	2	3	0	1
3	3	0	1	2

·	0	1	2	3
0	0	0	0	0
1	0	0	0	0
2	0	1	2	3
3	0	1	2	3

Figure 6.4

Exercise 6.1

1. Using the notational conventions introduced in this section, write out in detail the axioms for a ring without using the group concept.

2. Consider the table below. For the set {a, b, c, d}, can this table be used to define an addition operation for a ring?

	a	b	c	d
a	a	b	c	a
b	b	c	c	d
c	c	d	a	b
d	a	a	b	c

3. Give the addition and multiplication tables for the ring $(Z_5, \oplus, \odot)$.

4. Show that addition is not distributive over multiplication in $(Z, +, \cdot)$.

5. Consider the set of rational numbers Q. We define addition $\oplus$ and multiplication $\odot$ on Q by $a \oplus b = ab$

and $a \odot b = a + b$ for all $a, b \in Q$, where $+$ and $\cdot$ are ordinary addition and multiplication on rational numbers. Is $(Q, \oplus, \odot)$ a ring?

6. Let $A = \{0, 1\}$ and define addition and multiplication on A as indicated in the following tables. Is $(A, +, \cdot)$ a ring?

+	0	1
0	0	1
1	1	0

·	0	1
0	0	1
1	0	1

7. Let $A = \{a, b, c\}$. The addition table and part of the multiplication table for the ring $(A, +, \cdot)$ are displayed below. Using the distributive laws, fill in the rest of the multiplication table. Is $(A, +, \cdot)$ a commutative ring with an identity?

+	a	b	c
a	a	b	c
b	b	c	a
c	c	a	b

·	a	b	c
a	a	a	a
b	a		
c	a		b

8. Let $A = Z \times Z$. We define addition and multiplication as follows:

$$\text{addition—}(a,b) + (c,d) = (a + c, b + d)$$
$$\text{multiplication—}(a,b)(c,d) = (ac, bd)$$

Is $(A, +, \cdot)$ a ring?

9. Show that the left distributive law fails in the system displayed in Figure 6.4.

10. Let A be a set and $\mathcal{P}(A)$ be the power set of A. For $X, Y \in \mathcal{P}(A)$ define $+$ and $\cdot$ as follows.

$$X + Y = (X \cup Y) - (X \cap Y)$$
$$X \cdot Y = X \cap Y$$

Show that $(\mathcal{P}(A), +, \cdot)$ is a ring. Is it a commutative ring with identity?

11. Recall the Abelian group of rigid motions of a rectangle, (S, ◇) (page 239). Let ◇ be the addition operation on S, and let the multiplication operation be defined on S, as indicated in the following table.

·	I	R	H	V
I	I	R	I	R
R	I	R	R	I
H	I	R	H	V
V	I	R	V	H

Is $(S, +, \cdot)$ a ring?

12. In Figure 6.3, a system is displayed in which it is true that $x \cdot 0 = 0$ for all $x \in A$. Is it true that $0 \cdot x = 0$ for all $x \in A$?

Section 2
Basic Theorems about Rings

We will now bring the machinery from past mathematical activities to bear on the deductive mathematical system called a ring. In the following theorems, we will prove basic facts about rings. Notice the strong reliance upon the distributive laws in the proofs of these theorems.

Theorem 1: Let $(R, +, \cdot)$ be a ring. If 0 is the additive identity of R, then for all $a \in R$, $a \cdot 0 = 0$.

Proof: Since $0 = 0 + 0$, we know that $a \cdot 0 = a(0 + 0)$. By the left distributive law, then, $a \cdot 0 = a \cdot 0 + a \cdot 0$, but by the additive cancellation law for the group $(R, +)$, $a \cdot 0 = 0$.

Theorem 2: Let $(R, +, \cdot)$ be a ring. Then $a\,(-b) = -(ab)$.

Proof: From Theorem 1, we know that $0 = a \cdot 0$. Now, $0 = b + (-b)$ for any $b \in R$. Hence, we can write $0 = a \cdot 0 = a[b + (-b)] = ab + a(-b)$ by the left distributive law. In the additive group $(R, +)$, the unique inverse of ab is $-(ab)$ so that $-(ab) = a(-b)$.

Theorem 3: Let $(R, +, \cdot)$ be a ring. Then, $(-a)(-b) = ab$.

Proof: By the associative law for the additive group $(R, +)$, we can write $[ab + a(-b)] + (-a)(-b) = ab + [a(-b) + (-a)(-b)]$. By Theorem 2, we know that $ab + a(-b) = 0$ so that the left side of this equality becomes $(-a)(-b)$. So, $(-a)(-b) = ab + [a(-b) + (-a)(-b)]$. Using the right distributive law on the right side of this equality, we have,

$(-a)(-b) = ab + [a + (-a)](-b)$. Hence, $(-a)(-b) = ab + 0 \cdot (-b) = ab + 0 = ab$. (Note that $0 \cdot a = 0$ for all $a \in R$ will be proved in the exercises.)

Perhaps some of you have had an experience similar to that of the authors. We recall that in our early student years the rules for multiplying and adding signed numbers seemed a bit magical. In fact, in secondary school algebra, it was still disturbing to be told that the product of two negative numbers was positive. As has been indicated, the number systems $(Z, +, \cdot)$, $(Q, +, \cdot)$, and $(R, +, \cdot)$ are all rings and hence obey the rules established by the previous theorems. We hope that realizing the source of these "truths" will put your minds at ease.

Exercises 6.2

1. Prove that $0 \cdot a = 0$ for all $a \in R$ in a ring $(R, +, \cdot)$.

2. Prove that $(-a)(b) = a(-b)$ in a ring $(R, +, \cdot)$.

3. Prove that $-(a + b) = -a - b$ in a ring $(R, +, \cdot)$.

4. Prove that $a(b - c) = ab - ac$ in a ring $(R, +, \cdot)$.

5. Prove that, if $(R, +, \cdot)$ is a ring with identity 1, then 1 is unique.

6. Let $(R, +, \cdot)$ be a ring such that $a^2 = a$ for all $a \in R$. (This is called a Boolean ring.) Prove that:
a) $a = -a$ for all $a \in R$.
b) $ab = ba$ for all $a, b \in R$.

7. Let $(R, +, \cdot)$ be a ring with identity 1. Prove that, if a is an element of R with a multiplicative inverse a^{-1}, then a^{-1} is unique.

8. Let $(R, +, \cdot)$ be a Boolean ring. Prove that the equation $a + x = b$ has the unique solution $x = a + b$.

9. Let $(R, +, \cdot)$ be a ring with an identity. Prove that, if $a, b \in R$ are where a^{-1} and b^{-1} are multiplicative inverses of a and b, respectively, then $(ab)^{-1} = b^{-1}a^{-1}$.

10. Let $(R, +, \cdot)$ be a ring with identity 1 such that R has at least two distinct elements. Prove that $0 \neq 1$.

Section 3
Subrings

The concept of a subring of a ring is analogous to that of a subgroup for a group. For example, the ring $(\{0, 2\}, \oplus, \odot)$ is a subring of $(Z_4, \oplus, \odot)$, and the ring $(2Z, +, \cdot)$, is a subring of $(Z, +, \cdot)$.

Definition: Let $(R, +, \cdot)$ be a ring. A nonempty subset R' of R is a *subring* of $(R, +, \cdot)$ if and only if $(R', +, \cdot)$ is a ring.

Notice that $R' \subset R$ is a subring if and only if it is a ring with respect to binary operations defined on R.

As with groups, the following characterization theorem is sometimes useful in showing that a subset of a given ring is in fact a subring of the given ring.

Theorem 4: Let $(R, +, \cdot)$ be a ring such that R' is a nonempty subset of R. $(R', +, \cdot)$ is a subring of $(R, +, \cdot)$ if and only if $a - b$ and ab are in R' for all $a, b \in R'$.

Proof: *Part I:* If $(R', +, \cdot)$ is a subring of $(R, +, \cdot)$, then $(R', +, \cdot)$ is a ring; hence, if $a, b \in R'$ then $a - b$ and ab are in R'.

Part II: If $R' \neq \phi$ and contains $a - b$ for all $a, b \in R'$, then $(R', +)$ is a subgroup of $(R, +)$ according to Theorem II, Chapter 5. Since the distributive property holds for all elements of R, it holds for all of the elements of R'. Similarly, associativity holds in R'. Since $ab \in R'$ for all $a, b \in R'$, $\cdot$ is a binary operation on R'. Hence, $(R', +, \cdot)$ is a ring.

Exercises 6.3

1. Find two subrings of $(Z, +, \cdot)$.

2. Find three subrings of $(Z_{12}, \oplus, \odot)$.

3. Let $(R, +, \cdot)$ be a ring, and let $(R', +, \cdot)$ and $(R'', +, \cdot)$ be subrings of $(R, +, \cdot)$. Prove that $(R' \cap R'', +, \cdot)$ is a subring of $(R, +, \cdot)$.

4. Let $(R, +, \cdot)$ be a ring, and let a be a fixed element of R. Define $R_a = \{x \in R \mid ax = 0\}$. Prove that $(R_a, +, \cdot)$ is a subring of $(R, +, \cdot)$.

5. Consider the ring $(Z_{12}, \oplus, \odot)$. What is $(R_2, +, \cdot)$ for R_2 defined as in Problem 4?

6. Let $(R, +, \cdot)$ be a ring. Prove that, if $(R', +, \cdot)$ is a subring of $(R, +, \cdot)$, then $(R', +)$ is a subgroup of the additive group $(R, +)$.

Section 4
Integral Domains

Think of the ring $(Z_6, \oplus, \odot)$. This is a commutative ring with an identity element. Notice that $3 \odot 2 = 0$. Hence, in this ring, it is the case that we have two nonzero elements whose product is 0. The ring $(\{0, 1\}, +, \cdot)$, defined by the multiplicative and additive tables in Figure 6.5, is another example of a ring that has two nonzero elements whose product is 0.

+	0	1
0	0	1
1	1	0

·	0	1
0	0	0
1	0	0

Figure 6.5

Definition: Let $(R, +, \cdot)$ be a ring. a and b elements of R are *divisors of zero* if and only if $a \neq 0$ and $b \neq 0$ and $ab = 0$.

The rings $(Z_4, \oplus, \odot)$ and $(\{0, 1\}, +, \cdot)$ are rings with divisors of zero. The ring $(Z_3, \oplus, \odot)$ does not have any divisors of zero. Recall that $(Z_m, \oplus, \odot)$ is a ring. It is clear that $(Z_m, \oplus, \odot)$ is a ring with divisors of zero if and only if m is a composite number.

Definition: An *integral domain* is a commutative ring with identity, having at least two elements and no divisors of zero.

We require that an integral domain have at least two elements to rule out the system consisting of zero alone, which has no zero divisors by default.

The ring $(Z, +, \cdot)$ is a familiar example of an integral domain.

Definition: A ring $(R, +, \cdot)$ satisfies the cancellation law for multiplication if and only if

a) when $ac = bc$, then $a = b$ (right cancellation),

b) and when $ca = cb$, then $a = b$ (left cancellation),

where a, b, c are elements of R and $c \neq 0$.

Notice in the commutative ring with identity, $(Z_{12}, \oplus, \odot)$, that $3 \odot 4 = 0$ and $3 \odot 0 = 0$ so that $3 \odot 4 = 3 \odot 0$ and yet $4 \neq 0$. Hence, the cancellation laws for multiplication do not hold in $(Z_{12}, \oplus, \odot)$.

The integral domain $(Z, +, \cdot)$ is an example of a ring in which the cancellation laws hold for multiplication.

The following theorem along with Problem 1 of Exercises 6.4 guarantees us that the cancellation laws hold in any integral domain.

Theorem 5: Let $(R, +, \cdot)$ be an integral domain. If a, b, c are elements of R such that $a \neq 0$ and $ab = ac$, then $b = c$.

Proof: Since $ab = ac$, we know that $ab - ac = ac - ac = 0$. But $ab - ac = a(b - c)$ from Problem 4, Exercises 6.2. Since $a \neq 0$ by hypothesis and $(R, +, \cdot)$ an integral domain means that when a product is equal to 0 then one of the factors in the product must be 0. We know that $b - c$ must equal 0.

Exercises 6.4

1. Let $(R, +, \cdot)$ be an integral domain. Prove that, if $a, b, c \in R$ such that $a \neq 0$ and $ba = ca$, then $b = c$.

2. Find all of the divisors of zero in each of the following rings.

a) $(Z_6, \oplus, \odot)$ c) $(Z, +, \cdot)$

b) $(Z_7, \oplus, \odot)$ d) $(Z_{12}, \oplus, \odot)$

3. Let $(R, +, \cdot)$ be a commutative ring with identity. Prove that, if the cancellation laws hold for all elements of R, then $(R, +, \cdot)$ is an integral domain.

4. Let $(R, +, \cdot)$ be an integral domain. Prove that, if $(R', +, \cdot)$ is a subring of $(R, +, \cdot)$ such that $1 \in R'$, then $(R', +, \cdot)$ is an integral domain.

Section 5
Fields

We now consider those commutative rings with identity, which, unlike the integers, include a multiplicative inverse for every nonzero element of the ring. The familiar number systems $(\mathrm{Q}, +, \cdot)$ and $(\mathbf{R}, +, \cdot)$ are examples of such rings.

Definition: A *field* is a commutative ring with identity having at least two elements such that every nonzero element has a multiplicative inverse.

As with an integral domain, we require that a field have at least two elements to rule out the system consisting of 0 alone, which would satisfy the multiplicative inverse condition vacuously.

Theorem 6: Every field is an integral domain.

Proof: Let $(R, +, \cdot)$ be a field. We need only show that R has no divisors of zero since all of the other properties for an integral domain are given in the definition of a field. To do this, we suppose $a, b \in R$ such that $a \cdot b = 0$ and $a \neq 0$. We must now show that $b = 0$. Since $a \neq 0$, this means there exists an $a^{-1} \in R$ such that $a^{-1}a = 1$. Multiplying both sides of $ab = 0$ by a^{-1}, we have $a^{-1}(ab) = a^{-1} \cdot 0 = 0$. But $a^{-1}(ab) = (a^{-1}a)b = 1 \cdot b = b$ so that $b = 0$.

Since a field is also an integral domain, the cancellation

laws hold in a field. Experience with the fields $(Q, +, \cdot)$ and $(\mathbf{R}, +, \cdot)$ makes this a familiar fact.

There is a condition under which fields and integral domains are equivalent concepts. This surprising and interesting result is stated in the next theorem.

Theorem 7: Every finite integral domain is a field.

Proof: Let $(R, +, \cdot)$ be a finite integral domain. We need to show that for every nonzero element a of R, that there exists b, element of R, such that $ab = 1$. We display the elements of R in the following way: $0, a_1, a_2, a_3, \ldots, a_n$. We let a be one of the a_i (where $i \in \{1, 2, 3, \ldots, n\}$) in the list of elements of R. Now we multiply each element in the preceeding list by a. Hence, we have $a \cdot 0, aa_1, aa_2, aa_3, \ldots, aa_n$. Now, the only zero in this list is $a \cdot 0$ since an integral domain has no zero divisors. Also, each element in the list $aa_1, aa_2, aa_3, \ldots, aa_n$ is distinct since for any aa_j and aa_i (where $i, j \in \{1, 2, 3, \ldots, n\}$) in the list, $aa_i = aa_j$ implies $a_i = a_j$ since the cancellation laws hold in an integral domain. Hence, $a \cdot 0, aa_1, aa_2, aa_3, \ldots, aa_n$ are exactly the elements $0, a_1, a_2, a_3, \ldots, a_n$ in some order. Thus, $aa_i = 1$ for some $a_i \in R$ since an integral domain has a multiplicative identity. Hence, this a_i is the b we are looking for. Thus, a has a multiplicative inverse.

From this theorem, we realize that since $(Z_m, \oplus, \odot)$ is an integral domain if m is a prime, then $(Z_m, \oplus, \odot)$ is also a field. These are probably the first examples of finite fields you have encountered.

The diagram in Figure 6.6 helps to organize the relationship of the various rings we have been discussing.

Exercises 6.5

1. Which of the following are fields?

a) $(Z_2, \oplus, \odot)$ d) $(Z_4, \oplus, \odot)$
b) $(Z_3, \oplus, \odot)$ e) $(Z_6, \oplus, \odot)$
c) $(Z, +, \cdot)$

2. Define a field without using the concept of a group or a ring.

3. Let $(R, +, \cdot)$ be a field. Prove that, if $(R', +, \cdot)$ is a

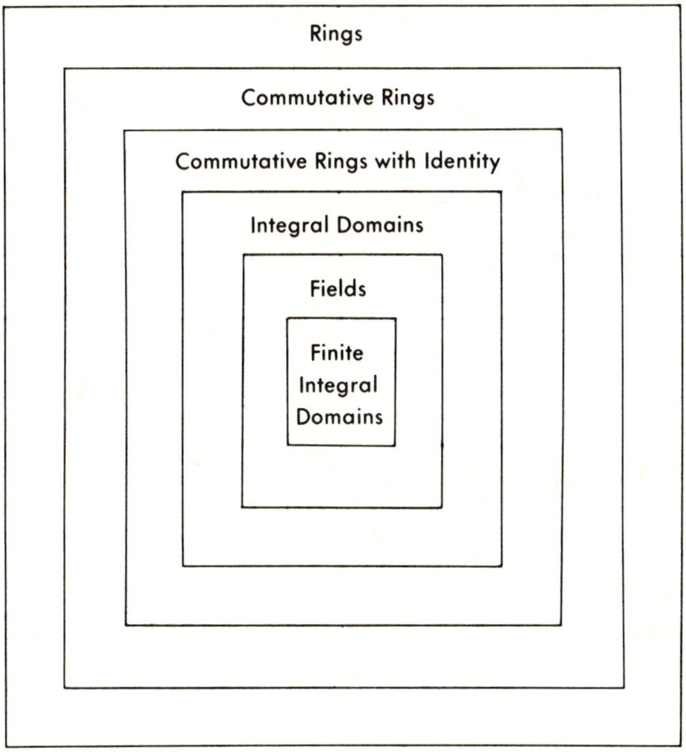

Figure 6.6

subring of $(R,+,\cdot)$ such that $1 \in R'$, then $(R',+,\cdot)$ is an integral domain.

4. Let $(R,+,\cdot)$ be a field. Let R' be a subset of R such that R' contains at least one nonzero element and such that for all a and b in R', it is true that $a + b$, $a - b$, and ab are in R' and for $b \neq 0$, that ab^{-1} is in R'. Prove that $(R',+,\cdot)$ is a field.

5. Let $(F,+,\cdot)$ be a field. Prove that if $a \in F$ such that $a \neq 0$, then $a \cdot x = b$ has a unique solution in $(F,+,\cdot)$.

6. Let $(R,+,\cdot)$ be a ring with identity. Let $G = \{u \mid u \in R$ such that u has a multiplicative inverse in $R\}$. Prove that $(G,\cdot)$ is a group.

7. For the rings $(Z,+,\cdot)$ and $(Z_{12},+,\cdot)$, illustrate Problem 6.

Section 6
Isomorphisms of a Ring

In the previous chapter, we discussed structurally identical groups and pointed out that such a structure concept was of basic importance. This same structure concept, as it relates to rings, is also fundamental. The notion of two rings being the same except for a relabeling of the names of their elements leads us, as was the case with groups, to make the following definition.

Definition: Let $(R, +, \cdot)$ and $(S, \oplus, \odot)$ be rings. A function f that is a one-to-one onto function from R to S is called a *ring isomorphism* if and only if $f(a + b) = f(a) \oplus f(b)$ and $f(ab) = f(a) \odot f(b)$ for every $a, b \in R$. $(R, +, \cdot)$ and $(S, \oplus, \odot)$ are said to be isomorphic, denoted $(R, +, \cdot) \cong (S, \oplus, \odot)$.

The theory of ring isomorphisms is similar to that of group isomorphisms. Notice that a function between rings is an isomorphism if and only if it preserves both operations.

Example 9: The ring $(Z_2, +, \cdot)$ is isomorphic to the ring $(A, \oplus, \odot)$, which is displayed in the tables in Figure 6.7. Let A =

$\oplus$	a	b
a	a	b
b	b	a

$\odot$	a	b
a	a	a
b	a	b

Figure 6.7

$\{a, b\}$. Let $f:Z_2 \to A$ be defined by $f(0) = a$ and $f(1) = b$. Then for the operations of $+$ and $\oplus$ we have the following.

a) $f(0 + 0) = f(0) = a$ and $f(0) \oplus f(0) = a \oplus a = a$ so that $f(0 + 0) = f(0) \oplus f(0)$.

b) $f(0 + 1) = f(1) = b$ and $f(0) \oplus f(1) = a \oplus b = b$ so that $f(0 + 1) = f(0) \oplus f(1)$.

c) $f(1 + 0) = f(1) = b$ and $f(1) \oplus f(0) = b \oplus a = b$ so that $f(1 + 0) = f(1) \oplus f(0)$.

d) $f(1 + 1) = f(0) = a$ and $f(1) \oplus f(1) = b \oplus b = a$ so that $f(1 + 1) = f(1) \oplus f(1)$.

Similarly, for $\cdot$ and $\odot$ we have,

a) $f(0 \cdot 0) = f(0) = a$ and $f(0) \odot f(0) = a \odot a = a$ so that $f(0 \cdot 0) = f(0) \odot f(0)$;

b) $f(0 \cdot 1) = f(0) = a$ and $f(0) \odot f(1) = a \odot b = a$ so that $f(0 \cdot 1) = f(0) \odot f(1)$;

c) $f(1 \cdot 0) = f(0) = a$ and $f(1) \odot f(0) = b \odot a = a$ so that $f(1 \cdot 0) = f(1) \odot f(0)$;

d) $f(1 \cdot 1) = f(1) = b$ and $f(1) \odot f(1) = b \odot b = b$ so that $f(1 \cdot 1) = f(1) \odot f(1)$.

Example 10: Recall that $(Z, +)$ and $(2Z, +)$ are isomorphic groups. Since the product of any two even integers is even, multiplication is a binary operation on $2Z$ so that $(2Z, +, \cdot)$ is a ring. Although $(Z, +)$ and $(2Z, +)$ are isomorphic under the isomorphism $f:Z \to 2Z$ as defined by $f(z) = 2z$ for all $z \in Z$, it is not the case that f is a ring isomorphism between Z and 2Z since $f(z_1 \cdot z_2) = 2(z_1 \cdot z_2)$ and $f(z_1) \cdot f(z_2) = (2z_1)(2z_2) = 4z_1z_2$ so that $f(z_1 \cdot z_2) \neq f(z_1) \cdot f(z_2)$.

The most basic properties of ring isomorphism are stated in the following theorems.

Theorem 8: Let $(R, +, \cdot)$ and $(S, +, \cdot)$ be rings. Let $f:R \to S$ be a ring isomorphism between these rings. If 0 is the zero of $(R, +, \cdot)$, then $f(0)$ is the zero of $(S, +, \cdot)$.

Proof: Consider the additive groups $(R, +)$ and $(S, +)$. From Theorem 14 (page 279) for group isomorphisms, $f(0)$ is the zero of $(S, +, \cdot)$ if 0 is the zero of $(R, +, \cdot)$.

Theorem 9: Let $(R, +, \cdot)$ and $(S, +, \cdot)$ be rings. Let $f:R \to S$ be a ring isomorphism between these rings. If a is any element of $(R, +, \cdot)$, then $f(-a) = -f(a)$.

Proof: Consider the additive groups $(R, +)$ and $(S, +)$. From Theorem 15 (page 279) for group isomorphisms, $f(-a) = -f(a)$ for any a of $(R, +, \cdot)$.

Theorem 10: Let $(R, +, \cdot)$ and $(S, +, \cdot)$ be rings. Let $f: R \rightarrow S$ be a ring isomorphism between these rings. If $(R, +, \cdot)$ is a commutative ring, then $(S, +, \cdot)$ is a commutative ring.

Proof: We must show that $xy = yx$ for all x, y elements of S. Since f is an isomorphism it is a 1–1, onto function. So, for x, y elements of S, there exist unique elements a, b of R such that $f(a) = x$ and $f(b) = y$. Now, $ab = ba$ since $(R, +, \cdot)$ is commutative and so $f(ab) = f(ba)$. Hence, we have $yx = f(b) \cdot f(a) = f(ba) = f(ab) = f(a) \cdot f(b) = xy$.

Exercises 6.6

1. Consider the ring $(A, +, \cdot)$, with addition and multiplication tables that follow, and the ring $(Z_3, \oplus, \odot)$.

+	a	b	c
a	a	b	c
b	b	c	a
c	c	a	b

·	a	b	c
a	a	a	a
b	a	b	c
c	a	c	b

Show an isomorphism between these rings.

2. Recall Problem 8 of Exercise 6.1. Consider all of the elements of $Z \times Z$ of the form $(a,0)$. Show that $A = \{(a,0) \mid a \in Z\}$ together with the defined operations of addition and multiplicaton is a subring of $Z \times Z$. Prove that this subring is isomorphic to $(Z, +, \cdot)$.

3. Show that $(2Z, +, \cdot)$ and $(5Z, +, \cdot)$ are not isomorphic rings.

4. Let $(R, +, \cdot)$ and $(S, +, \cdot)$ be rings with identity, and let $f: R \rightarrow S$ be a ring isomorphism between these rings. Prove that, if 1 is the identity of $(R, +, \cdot)$, then $f(1)$ is the identity of $(S, +, \cdot)$

5. Let $(R, +, \cdot)$ and $(S, +, \cdot)$ be rings and $f:R \to f(R)$ be a ring isomorphism from R to f(R) such that $f(R) \subset S$. Prove that f(R) is a subring of S.

6. Let $(R, +, \cdot)$ and $(S, +, \cdot)$ be rings and $f:R \to S$ be a ring isomorphism between these rings. Prove that, if a is an element of $(R, +, \cdot)$ having a multiplicative inverse, then $f(a^{-1}) = (f(a))^{-1}$.

7. Let $(R, +, \cdot)$ be a field and $(S, +, \cdot)$ be a ring. Prove that, if $f:R \to S$ is an isomorphism between $(R, +, \cdot)$ and $(S, +, \cdot)$, then $(S, +, \cdot)$ is a field.

Section 7
Concluding Remarks

The familiar number systems $(Z, +, \cdot)$, $(Q, +, \cdot)$, and $(\mathbf{R}, +, \cdot)$ are models of rings that we have been working with since early years. These structures are perhaps the most important models of rings as they relate to our daily experience. By considering the properties of rings abstractly, we have been able not only to look at these important models of rings but also to extend our awareness of mathematical ideas beyond these boundaries. Much time and effort on the part of the mathematician is spent in consideration not only of the familiar mathematical setting but also of areas completely outside the domain of everyday experience.

We hope the student has found a few pearls of knowledge and gained a keen appreciation for mathematics while studying this work and that the methods and processes of mathematics will be of benefit to him in whatever area of activity he pursues.

References
Chapter Six

Bell, E. T., *Men of Mathematics* (New York: Simon and Schuster, 1973), pp. 355–361.

Thompson, Robert C., and Yaqub, Adil, *Introduction to Linear and Abstract Algebra* (Glenview, Ill.: Scott, Foresman and Company, 1971), pp. 21–43.

Answers to Selected Exercises

Exercises 1.1

1. a) Well-defined
b) Not well-defined
c) Not well-defined
d) Well-defined
e) Not well-defined

3. a) True
b) False
c) True
d) True
e) False
f) True
g) False
h) False

5. a) {red, white, blue}
b) {January, February, March, April, May, June, July, August, September, October, November, December}
c) $\{6, 12, 18, 24, \ldots\}$
d) $\{14, 15, 16, 17, \ldots, 34, 35, 36\}$
e) $\{1, 2, 4, 8\}$
f) $\{a, i, l, u, m, n\}$

7. Answers vary. Possible answers are:
a) The set of natural numbers; $U = \{1, 2, 3, \ldots, 7\}$
b) The set of natural numbers; $U = A$
c) The set of all names of cities in California; U is the set of all names of cities in the U.S.
d) The set of all fish; U is the set of all animals.

9. a) Alabama, Arkansas, Arizona, Alaska
b) Alabama, Arizona, Alaska
c) Arizona, Alaska, Arkansas
d) Alaska

11. a) {1,3,4,5,6,7,9}
b) {3,7}
c) {4,6,1,5,9}

13. A = {{a, b}, {1, 2}}

15. a) Each element of A has exactly two vowels in it.
b) Each element is a member of the set A.

Exercises 1.2, 3

1.

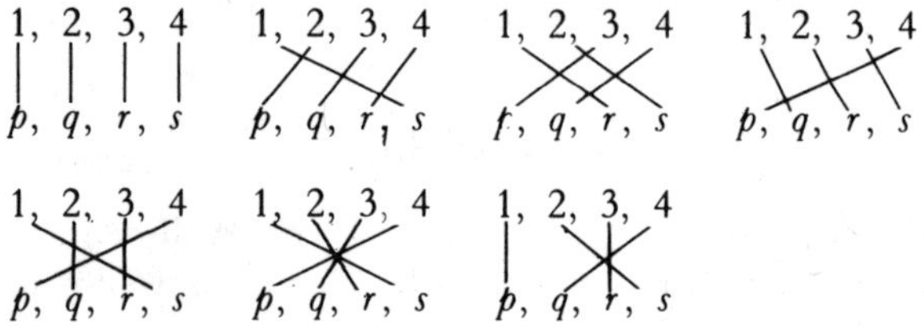

or other possibilities.

2. a) True d) False g) False j) False
b) True e) False h) False k) False
c) True f) True i) True l) True

3. (a), (d)

5. a) B = C
b) B ≡ C, C ≡ D, B ≡ D
c) B ⊂ C, C ⊂ B, A ⊂ D
d) A ⊂ D

7. a) False d) True g) False j) False
b) True e) True h) False
c) False f) True i) True

9. U, ϕ

10. a) 1
b) $1 \cdot 2 = 2$
c) $1 \cdot 2 \cdot 3 = 6$
d) $1 \cdot 2 \cdot 3 \cdot 4 = 24$
$1 \cdot 2 \cdot 3 \cdot \ldots \cdot n$ = number of one-to-one correspondences between two sets each having n elements.

13. a) A ≡ B c) A ≢ B e) A ≡ B
b) A ≡ B d) A ≢ B

15. a) True c) True e) True
b) False d) False f) True

17. ϕ, {a}, {{b, c}}, {a, {b, c}}

Exercises 1.4

1. a) Yes c) {4, 5, 6, 7, 8, 9, 11, 12, 13}
b) Yes d) {8, 12}

2. a) {1, 2, 8, 9, 10, 11, 12, 13}
c) {3, 5}
f) ϕ
k) {1, 2, 8, 10, 12, 13}
p) U
r) {4, 6, 7}

3. a) $A \not\subset C$ d) $A \cup B = U$
b) $B \subset C$ e) $B \cap \overline{C} = \phi$
c) $A \cap B = \phi$

5. $B \subset A$

7. $A = B$

9. $A = B = \phi$

11. $A = B$ or $B \subset A$

13. $A = B$

15. $A = \phi$

17. ϕ

18. $2^n + n$

19. a) $\mathcal{P}(A) = \{\phi, \{1\}, \{2\}, \{1,2\}\}$
b) $\mathcal{P}(\mathcal{P}(A)) = \{\phi, \{\phi\}, \{\{1\}\}, \{\{2\}\}, \{\{1,2\}\}, \{\phi,\{1\}\}, \{\phi,\{2\}\}, \{\phi,\{1,2\}\}, \{\{1\},\{2\}\}, \{\{1\},\{1,2\}\}, \{\{2\},\{1,2\}\}, \{\phi,\{1\}, \{2\}\}, \{\phi,\{1\}, \{1,2\}\}, \{\phi,\{2\}, \{1,2\}\}, \{\{1\}, \{2\}, \{1,2\}\}, \{\phi,\{1\}, \{2\}, \{1,2\}\}\}$

20. a) No c) Yes e) No
b) No d) Yes f) Yes

21. $\overline{A}$, A

23. a) {b, c, d}
b) {b, c, d}
c) ϕ
d) {f, g}
e) ϕ

24. {1, 2, 3, 5, 6, 7, 8}

Exercises 1.5

1\. a)

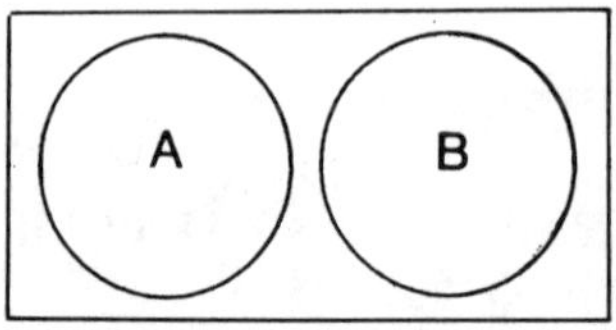

b)

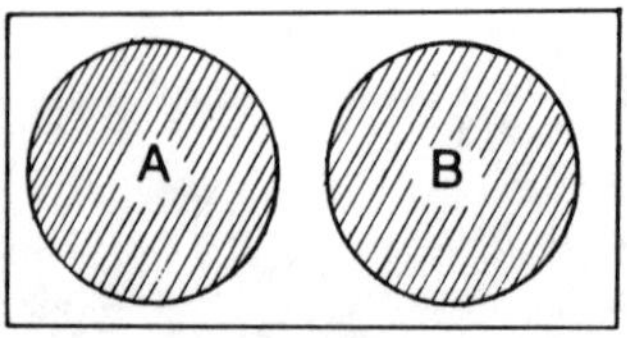

3\. a) $R \cap (S \cup T)$

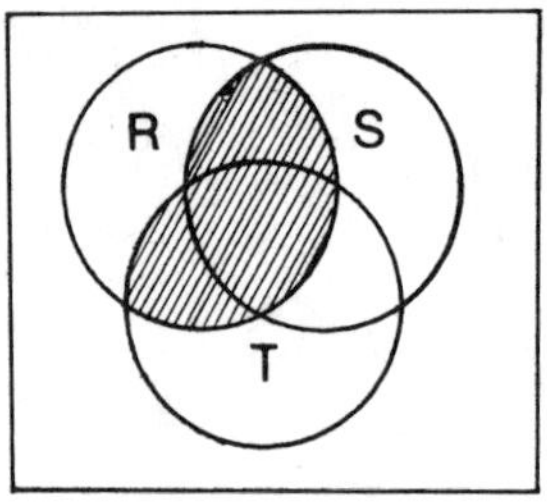

g) $(R \cap S) - T$

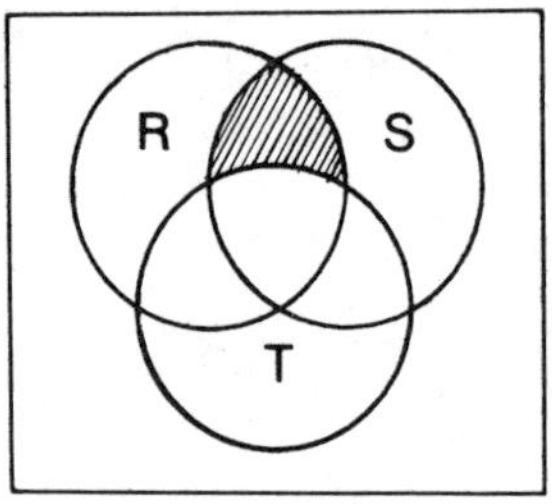

d) $(R \cup S) \cap (R \cup T)$

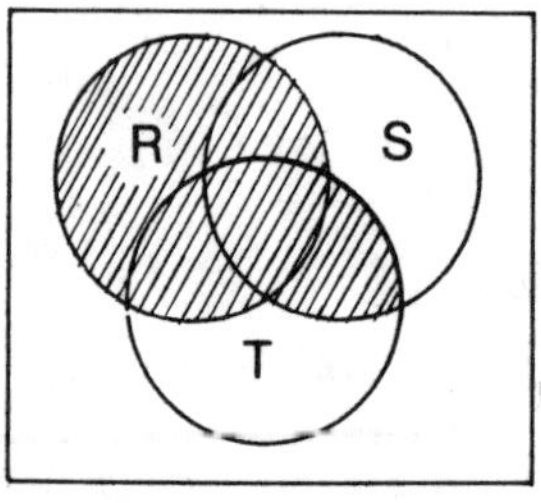

i) $[(R \cup S) - T] \cap R$

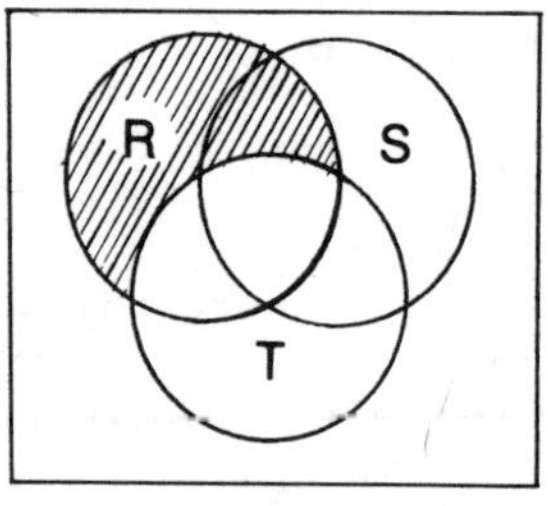

e) $(R \cap S) \cap T$

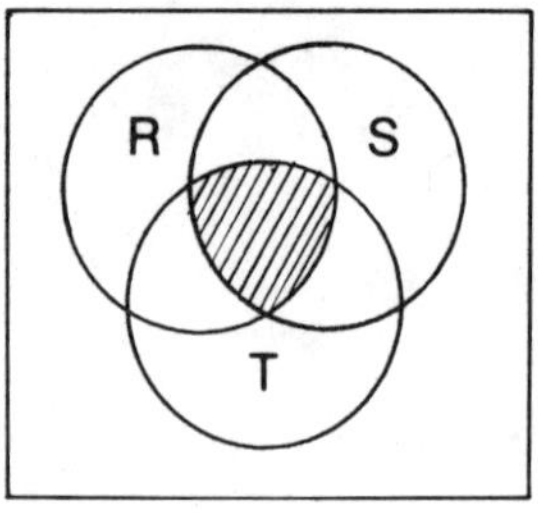

5. a)

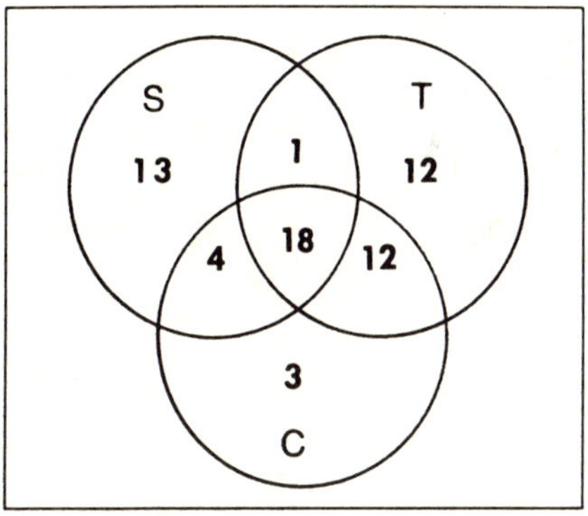

b) 12

c) 13

d) 67

7.

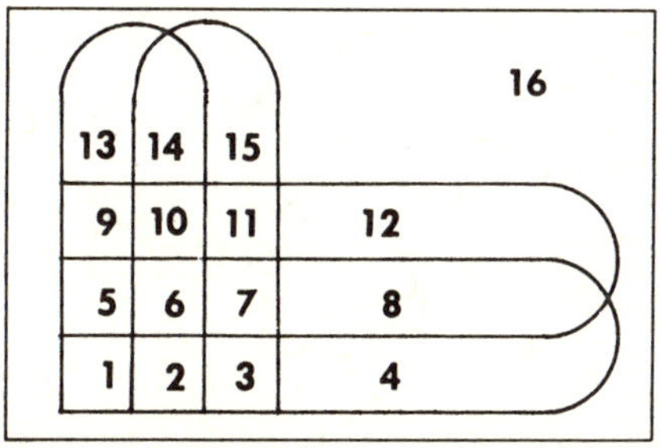

9. $A \, \Delta \, B = \{x \mid x \in (A \cap \overline{B}) \text{ or } x \in (B \cap \overline{A})\}$

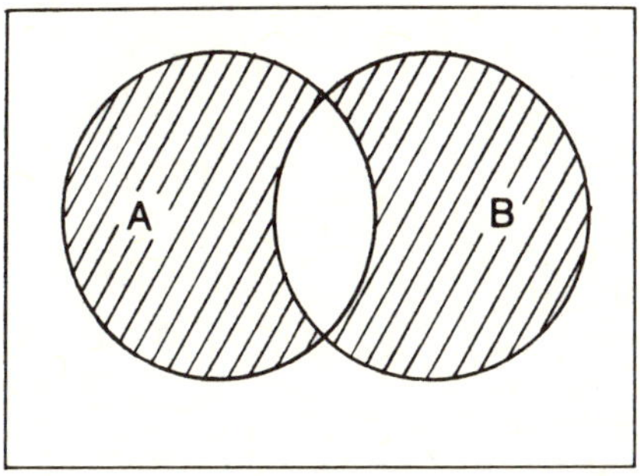

Exercises 1.6

1. a) A d) U g) ϕ j) ϕ m) A p) U
b) ϕ e) A h) U k) ϕ n) U q) ϕ
c) A f) A i) $\overline{A}$ l) ϕ o) ϕ r) U

3. a) $(S \cap R) \cup (S \cap T)$
b) $R \cup (S \cap T)$
c) $(S \cap R) \cup T$
d) $(A \cap C) \cup (B \cap C)$
e) $A \cap (B \cup C)$
f) $A \cup [(A \cup C) \cap (B \cup C)] = A \cup [(A \cap B) \cup C]$
g) $[R \cup (S \cap T)] \cap [R \cup (T \cap R)]$

4. a) B d) $B \cap \overline{A}$

5. a) $\overline{A \cup B}$ $\overline{A} \cap \overline{B}$

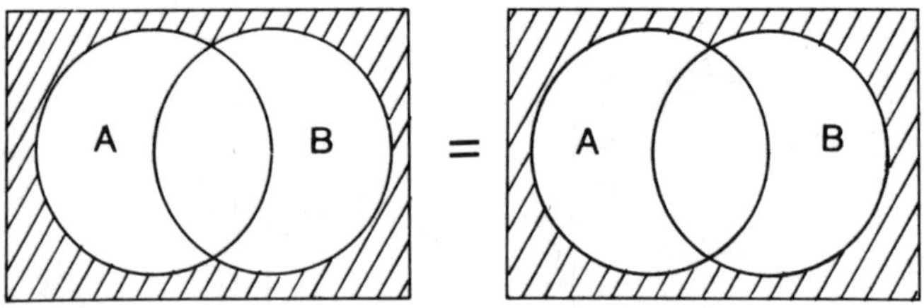

$\overline{A \cap B}$ $\overline{A} \cup \overline{B}$

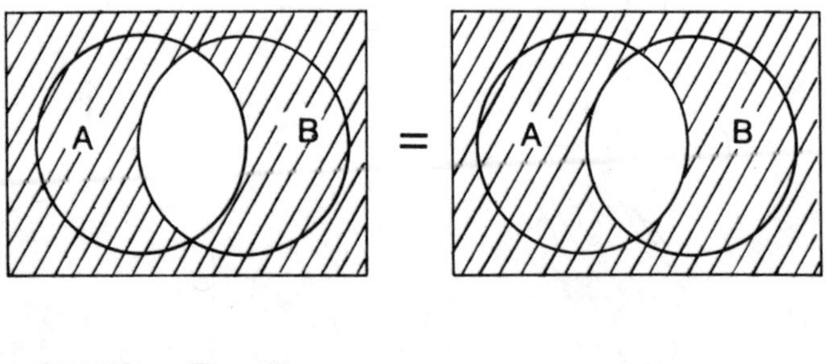

b) $\overline{A \cup B} = \overline{A} \cap \overline{B}$

Part 1: Let $x \in \overline{A \cup B}$. Then $x \notin A \cup B$. So, $x \notin A$ and $x \notin B$. Thus, $x \in \overline{A}$ and $x \in \overline{B}$, so $x \in \overline{A} \cap \overline{B}$. Hence, $\overline{A \cup B} \subset \overline{A} \cap \overline{B}$.

Part 2: Now, suppose $x \in \overline{A} \cap \overline{B}$. This means $x \in \overline{A}$ and $x \in \overline{B}$. So, $x \notin A$ and $x \notin B$. Thus, $x \notin A$ or B. Hence, $x \notin A \cup B$ and, so $x \in \overline{A \cup B}$. Hence, $(\overline{A} \cap \overline{B}) \subset \overline{A \cup B}$.

By Parts 1 and 2, $(\overline{A \cup B}) = \bar{A} \cap \bar{B}$

$\overline{A \cap B} = \bar{A} \cup \bar{B}$

Part 1: Let $x \in \overline{A \cap B}$. Then $x \notin A \cap B$. So either $x \notin A$ or $x \notin B$. Thus $x \in \bar{A}$ or $x \in \bar{B}$. Hence $x \in \bar{A} \cup \bar{B}$ and so $(\overline{A \cap B}) \subset \bar{A} \cup \bar{B}$.

Part 2: Let $x \in \bar{A} \cup \bar{B}$. Then $x \in \bar{A}$ or $x \in \bar{B}$. This means either $x \notin A$ or $x \notin B$. Thus, $x \notin A \cap B$. Hence, $x \in \overline{A \cap B}$, which means $\bar{A} \cup \bar{B} \subset (\overline{A \cap B})$.

By Parts 1 and 2 we have that $\overline{A \cap B} = \bar{A} \cup \bar{B}$.

c) $\overline{(A \cap B) \cap C} = (\bar{A} \cup \bar{B}) \cup \bar{C}$

$\overline{(A \cup B) \cup C} = (\bar{A} \cap \bar{B}) \cap \bar{C}$

7. $(A \cap B) \cap \{(C \cap D) \cap [(E \cap F) \cap G]\}$

$[(A \cap B) \cap C] \cap [(D \cap E) \cap (F \cap G)]$

$A \cap \{[(B \cap C) \cap (D \cap E)] \cap (F \cap G)\}$

$[A \cap (B \cap C)] \cap [(D \cap E) \cap (F \cap G)]$

$(A \cap B) \cap \{[(C \cap D) \cap E] \cap (F \cap G)\}$

$A \cap \{(B \cap C) \cap [(D \cap E) \cap (F \cap G)]\}$

$\{[(A \cap B) \cap (C \cap D)] \cap (E \cap F)\} \cap G$ or other possibilities.

9. b) $A - (B \cup C) = (A - B) \cap (A - C)$

Part 1: Show $A - (B \cup C) \subset (A - B) \cap (A - C)$

Let $x \in A - (B \cup C)$. Then $x \in A$ and $x \notin B \cup C$. Thus $x \in A$ and $x \notin B$ and $x \notin C$. So $x \in A$ and $x \notin B$ and $x \in A$ and $x \notin C$. Hence, $x \in A - B$ and $x \in A - C$. Thus $x \in (A - B) \cap (A - C)$. Hence, $A - (B \cup C) \subset (A - B) \cap (A - C)$.

Part 2: Show $(A - B) \cap (A - C) \subset A - (B \cup C)$

Let $x \in (A - B) \cap (A - C)$. Then $x \in A - B$ and $x \in A - C$. So, $x \in A$ and $x \notin B$ and $x \in A$ and $x \notin C$. Thus $x \in A$ and $x \notin B$ and $x \notin C$. So, $x \in A$ and $x \notin B$ or C. Hence, $x \in A$ and $x \notin B \cup C$. So $x \in A - (B \cup C)$. Thus $(A - B) \cap (A - C) \subset A - (B \cup C)$.

By Parts 1 and 2, $A - (B \cup C) = (A - B) \cap (A - C)$

d) $A - (B \cap C) = (A - B) \cup (A - C)$

Part 1: Let $x \in A - (B \cap C)$. Then $x \in A$ and $x \notin B \cap C$, so that $x \in A$ and either $x \notin B$ or $x \notin C$. Hence, $x \in A$ and $x \notin B$ or $x \in A$ and $x \notin C$, which means $x \in A$ and $x \in \bar{C}$ or $x \in A$ and $x \in \bar{B}$, and we have $x \in A \cap \bar{C}$ or $x \in A \cap \bar{B}$. By definition this means $x \in (A - C) \cup (A - B)$.

Part 2: Let $x \in (A - B) \cup (A - C)$. Then $x \in A - B$ or $x \in A - C$. So $x \in A$ and $x \notin B$ or $x \in A$ and $x \notin C$. In either case, $x \in A$. Thus, $x \in A$ and either $x \notin B$ or $x \notin C$. So, $x \in A$ and $x \notin B \cap C$. Hence, $x \in A \cap \overline{(B \cap C)}$. Thus, $x \in A - (B \cap C)$.

By Parts 1 and 2 we have, $A - (B \cap C) = (A - B) \cup (A - C)$

f) $A \cup (A \cap B) = A$

Part 1: Let $x \in A \cup (A \cap B)$. Then $x \in A$ or $x \in A \cap B$. *Case 1:* If $x \in A$, then $A \cup (A \cap B) \subset A$. *Case 2:* If $x \in A \cap B$, then $x \in A$ and $x \in B$, so that $x \in A$ and $A \cup (A \cap B) \subset A$. Hence, in either case $A \cup (A \cap B) \subset A$.

Part 2: Let $x \in A$. Then $x \in A \cup (A \cap B)$ so $A \subset A \cup (B \cap C)$

By Parts 1 and 2 we see that $A \cup (A \cap B) = A$.

Exercises 1.7

1. a) F
b) F
c) I
d) F
e) I
f) I
g) F
h) I
i) I
j) F

2. a) F
b) T
c) F
d) F
e) T
f) T
g) T
h) F
i) F

3. a) T
b) T
c) T
d) F
e) T
f) F
g) F
h) T

Exercises 1.8

1. a) 10
b) 26
c) 88
d) 9
e) 50
f) 333

2. a) C
b) O
c) C
d) O
e) C
f) O
g) C
h) O
i) O
j) O

3. $n(C) = n(T)$

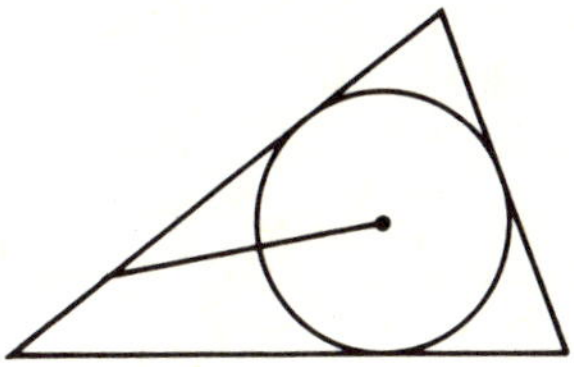

5. a) $A = \{1, 2, 3, \ldots\}$

$B = \{5, 10, 15, \ldots\}$

b) 995
c) $\aleph_0$

7. a) 3
b) 8
c) 256
d) 2^{256}

Number of elements in $\mathcal{P}(\mathcal{P}(A))$: 2^{2^k}

Exercises 1.9

1. a) $\{(p,5), (p,6), (p,7), (r,5), (r,6), (r,7), (s,5), (s,6), (s,7)\}$
b) $\{(p,r), (p,s), (p,p), (r,p), (r,s), (r,r), (s,p), (s,r), (s,s)\}$
c) $\{(5,p), (5,r), (5,s), (6,p), (6,r), (6,s), (7,p), (7,r), (7,s)\}$
d) $\{(5,5), (5,6), (5,7), (6,5), (6,6), (6,7), (7,5), (7,6), (7,7)\}$

3. a) $2 + 2 = 4$

Let $A = \{a,b\}$ and let $B = \{1,2\}$. Then, $n(A) = 2$, $n(B) = 2$, $A \cap B = \phi$, and $A \cup B = \{a, b, 1, 2\}$. So $2 + 2 = n(A) + n(B) = n(A \cup B) = 4$.

b) $3 + 5 = 8$

Let $A = \{1,2,3\}$ and let $B = \{o,p,q,r,s\}$. Then $A \cap B = \phi$ and $A \cup B = \{1,2,3,o,p,q,r,s\}$. So $n(A) = 3$, $n(B) = 5$, and $n(A \cup B) = 8$. Thus, $3 + 5 = n(A) + n(B) = n(A \cup B) = 8$.

c) $4 + 0 = 4$

Let $A = \{1,2,3,4\}$ and let $B = \phi$. Then, $A \cap B = \phi$, $A \cup B = \{1,2,3,4\}$, $n(A) = 4$, $n(B) = 0$, and $n(A \cup B) =$

4. So $4 + 0 = n(A) + n(B) = n(A \cup B) = 4$.

h) $0 + 0 = 0$

Let $A = \phi$, and let $B = \phi$. Then, $A \cap B = \phi$ and $A \cup B = \phi$. So since $n(A) = 0$, $n(B) = 0$, and $n(A \cup B) = 0$, then $0 + 0 = n(A) + n(B) = n(A \cup B) = 0$.

4. a) Let $A = \{a\}$, and let $B = \phi$. Then, $n(A) = 1$ and $n(B) = 0$. Then, $A \times B = \phi$. So, $n(A \times B) = 0$. Hence, $1 \cdot 0 = n(A) \cdot n(B) = n(A \times B) = 0$.

e) Let $A = \{r,s,t\}$, and let $B = \{\$\}$. Then, $n(A) = 3$ and $n(B) = 1$. Now, $A \times B = \{(r,\$),(s,\$),(t,\$)\}$. So that $n(A \times B) = 3$. Hence, $3 \cdot 1 = n(A) \cdot n(B) = n(A \times B) = 3$.

5. $A = B$

7. a) $A \times (B \cup C) = (A \times B) \cup (A \times C)$

$A \times (B \cup C) = \{2,4\} \times \{a,e,i,o,u\} = \{(2,a), (2,e), (2,i), (2,o), (2,u), (4,a), (4,e), (4,i), (4,o), (4,u)\}$

$(A \times B) \cup (A \times C) = \{(2,a), (2,e), (2,i), (4,a), (4,e), (4,i)\} \cup \{(2,o), (2,u), (4,o), (4,u)\} = \{(2,a), (2,e), (2,i), (2,o), (2,u), (4,a), (4,e), (4,i), (4,o), (4,u)\}$

Thus, $A \times (B \cup C) = (A \times B) \cup (A \times C)$

b) $A \times (B \cap C) = (A \times B) \cap (A \times C)$

$A \times (B \cap C) = \{2,4\} \times \phi = \phi$

$(A \times B) \cap (A \times C) = (\{2,4\} \times \{a,e,i\}) \cap (\{2,4\} \times \{o,u\}) = \{(2,a,), (2,e), (2,i), (4,a), (4,e), (4,i)\} \cap \{(2,o), (2,u), (4,o), (4,u)\} = \phi$

Thus, $A \times (B \cap C) = (A \times B) \cap (A \times C)$

c) $A \times B = (A \times U) \cap (U \times B)$

$A \times B = \{2,4\} \times \{a,e,i\} = \{(2,a), (2,e), (2,i), (4,a), (4,e), (4,i)\}$

$(A \times U) \cap (U \times B) = \{(2,a), (2,e), (2,i), (2,o), (2,u), (2,2), (2,4), (4,a), (4,e), (4,i), (4,o), (4,u), (4,2), (4,4)\} \cap \{(a,a), (a,e), (a,i), (e,a), (e,e), (e,i), (i,a), (i,e), (i,i), (o,a), (o,e), (o,i), (u,a), (u,e), (u,i), (2,a), (2,e), (2,i), (4,a), (4,e), (4,i)\} = \{(2,a), (2,e), (2,i), (4,a), (4,e), (4,i)\}$

Thus, $A \times B = (A \times U) \cap (U \times B)$

d) $A \times (B - C) = (A \times B) \cap (A \times \overline{C})$

$A \times (B - C) = \{2,4\} \times \{a,e,i\} = \{(2,a), (2,e), (2,i), (4,a), (4,e), (4,i)\}$

$(A \times B) \cap (A \times \overline{C}) = \{(2,a), (2,e), (2,i), (4,a), (4,e), (4,i)\} \cap \{(2,a), (2,e), (2,i), (2,2), (2,4), (4,a), (4,e), (4,i), (4,2), (4,4)\} = \{(2,a), (2,e), (2,i), (4,a), (4,e), (4,i)\}$

Thus, $A \times (B - C) = (A \times B) \cap (A \times \overline{C})$

9. a) Let $(x,y) \in A \times (B \cup C)$. Then $x \in A$ and $y \in B \cup C$. Suppose $y \in B$. Then $x \in A$ and $y \in B$ means $(x,y) \in A \times B$ and so $(x,y) \in (A \times B) \cup (A \times C)$. Suppose $y \notin B$. Then $y \in C$ and so $(x,y) \in A \times C$ and so $(x,y) \in (A \times B) \cup (A \times C)$. In either case, $A \times (B \cup C) \subset (A \times B) \cup (A \times C)$. Let $(x,y) \in (A \times B) \cup (A \times C)$. Then $(x,y) \in A \times B$ or $(x,y) \in A \times C$. Suppose $(x,y) \in A \times B$. Then $x \in A$ and $y \in B$ so that $y \in B \cup C$ and so $(x,y) \in A \times (B \cup C)$. Suppose $(x,y) \notin A \times B$. Then $(x,y) \in A \times C$ and $x \in A$ and $y \in C$ so that $y \in B \cup C$ and so $(x,y) \in A \times (B \cup C)$. In either case, $(A \times B) \cup (A \times C) \subset A \times (B \cup C)$.

10. a) $\aleph_0$

b) 3

c) Yes

d) $A \cup B = \{@, \#, \$, 1, 2, 3, 4, \ldots\}$

$A = \{1, 2, 3, 4, 5, 6, 7, \ldots\}$ (elements paired one-to-one by vertical bars)

e) $\aleph_0$

f) $\aleph_0$

Exercises 2.1

1. a) Yes
b) No
c) Yes
d) No
e) Yes
f) No
g) No
h) No
i) Yes
j) Yes
k) Yes
l) Yes
m) No

2. Yes

3. a) Hurry up!
b) Come here.
c) Would you like a coke?
d) What a horrible movie!
e) I can't take this any more.
f) This sentence is false.

5. a) Yes
b) Depends on your accepted theory of cosmogony.

6. a) C
b) C
c) C
d) S
e) S
f) C
g) C
h) S
i) C
j) S
k) S

Exercises 2.2

1. a) S
b) N
c) S
d) S
e) N
f) S
g) O
h) S
i) O
j) O
k) N
l) S

3. a) "Red is a color in the U.S. flag" T
b) "Yellow is a color in the U.S. flag" F
c) "Blue is a color in the U.S. flag" T
d) "Black is a color in the U.S. flag" F

5. a) (1) "Houston is the name of a city with a population of more than 400,000 people." True
(2) "Milwaukee is the name of a city with a population of more than 400,000 people." True

b) (1) "Reno is the name of a city with a population of more than 400,000 people." False
(2) "Tucson is the name of a city with a population of more than 400,000 people." False

c) U = {Dallas, New York, Portland}
U = {Springfield, Little Rock, Pierre, Tombstone}

7. a) U = {violin, piano, flute, table, chair}
"Piano is a musical instrument"

b) U = {violin, table, chair, trumpet}
$\exists\, x \in U, P(x)$

c) U = {trumpet, piano, harp, flute, clarinet}
$\forall\, x \in U, P(x)$

8. a) T
b) F
c) F
d) T
e) T
f) T
g) F
h) T
i) T

9. a) U = {oranges, apples, pears}
U = {oranges, grapefruit, grapes, cherries}
U = {tangerines, grapefruit, pears}

b) $\forall x \in U, \overline{P(x)}$

Exercises 2.3, 4

1. Logically true: All cars are cars. Logically false: Today is not one of the seven days of the week.

2. a) It is not the case that this is not a very elaborate building. (This is a very elaborate building.)
c) The fence is not white.
g) It is not the case that we are going to the drugstore and then we are going to the ice rink.
j) It is not the case that Oliver is a dog or else he is a cat.

3. a) Every record club is a money making enterprise.
b) There exists a record club that is a money-making enterprise.
c) There exists a record club that is not a money making enterprise.
d) Every record club is not a money making enterprise.
e) It is not the case that every record club is a money making enterprise.
f) It is not the case that there exists a record club that is a money making enterprise.
g) It is not the case that there exists a record club that is not a money making enterprise.
h) It is not the case that every record club is not a money making enterprise.

5. a) $\forall x \in U, P(x)$ — I own a dress that does not need to have its hem raised.
c) $\exists x \in U, \overline{P(x)}$ — Every watermelon was ripe.
e) $\exists x \in U, P(x)$ — All good novels are not being written by current American authors.
g) $\forall x \in U, \overline{P(x)}$ — At least one applicant was well qualified.

Exercises 2.5, 6

1. a) No
b) No
c) Yes
d) Yes
e) No
f) No
g) Yes

2. a) I
   b) E
   c) I
   d) I
   e) E
3. a) $P \wedge Q$
   b) $Q \wedge \overline{P}$
   c) $P \vee Q$
   d) $Q \wedge P$
   e) $\overline{Q} \wedge P$
   f) $\overline{P \vee Q}$
   g) $\overline{\overline{P} \wedge \overline{Q}}$
   h) $\overline{P} \wedge \overline{Q}$
5. a) F
   b) F
   c) T
   d) F
   e) T
   f) T
   g) T

Exercises 2.7

1. a)

p	q	r	$p \wedge q$	$\overline{r}$	$(p \wedge q) \vee \overline{r}$
T	T	T	T	F	T
T	T	F	T	T	T
T	F	T	F	F	F
T	F	F	F	T	T
F	T	T	F	F	F
F	T	F	F	T	T
F	F	T	F	F	F
F	F	F	F	T	T

b)

p	q	r	$p \vee q$	$p \wedge r$	$(p \vee q) \wedge (p \wedge r)$
T	T	T	T	T	T
T	T	F	T	F	F
T	F	T	T	T	T
T	F	F	T	F	F
F	T	T	T	F	F
F	T	F	T	F	F
F	F	T	F	F	F
F	F	F	F	F	F

c)

p	q	$\bar{q}$	$p \wedge \bar{q}$
T	T	F	F
T	F	T	T
F	T	F	F
F	F	T	F

g)

p	q	r	$\bar{p}$	$\bar{p} \vee r$	$\bar{q}$	$\bar{q} \wedge (\bar{p} \vee r)$
T	T	T	F	T	F	F
T	T	F	F	F	F	F
T	F	T	F	T	T	T
T	F	F	F	F	T	F
F	T	T	T	T	F	F
F	T	F	T	T	F	F
F	F	T	T	T	T	T
F	F	F	T	T	T	T

k)

p	q	r	$\bar{p}$	$q \wedge \bar{p}$	$r \wedge \bar{p}$	$(q \wedge \bar{p}) \vee (r \wedge \bar{p})$	$[(q \wedge \bar{p}) \vee (r \wedge \bar{p})] \vee p$
T	T	T	F	F	F	F	T
T	T	F	F	F	F	F	T
T	F	T	F	F	F	F	T
T	F	F	F	F	F	F	T
F	T	T	T	T	T	T	T
F	T	F	T	T	F	T	T
F	F	T	T	F	T	T	T
F	F	F	T	F	F	F	F

n)

p	q	r	$\bar{p}$	$p \wedge r$	$\overline{p \wedge r}$	N $q \wedge (\overline{p \wedge r})$	$\bar{p} \vee$ N	$r \vee q$	$(r \vee q) \wedge$ $(\bar{p} \vee$ N$)$
T	T	T	F	T	F	F	F	T	F
T	T	F	F	F	T	T	T	T	T
T	F	T	F	T	F	F	F	T	F
T	F	F	F	F	T	F	F	F	F
F	T	T	T	F	T	T	T	T	T
F	T	F	T	F	T	T	T	T	T
F	F	T	T	F	T	F	T	T	T
F	F	F	T	F	T	F	T	F	F

2. a) $(p \wedge q) \wedge r$ e) $(\bar{p} \wedge \bar{q}) \vee (r \wedge s)$
c) $(p \vee q) \wedge r$ g) $(p \wedge q) \wedge (r \vee s)$

3. a) F d) T
b) F e) F
c) F

5.

p	q	$p \wedge q$	$\overline{p \wedge q}$	$p \vee q$	$(p \vee q) \wedge (\overline{p \wedge q})$
T	T	T	F	T	F
T	F	F	T	T	T
F	T	F	T	T	T
F	F	F	T	F	F

7. a) 8 c) 32
b) 16 d) 64

Exercises 2.8

1. U

3. a) F c) T e) F
b) F d) F f) F

4. a) CSF g) CSF
c) CSF i) T
e) T k) CSF

Exercises 2.9

1. a) It is not the case that fire destroys London and alligators have teeth.

b) It is not the case that Lincoln was famous and Babe Ruth did not play baseball.

c) It is not the case that spiders are creepy and roaches are not ugly.

d) It is not the case that the circus will come to town within the next two weeks and I will not go to a performance.

2. a) If I get to my math class on time, then I must leave my home fifteen minutes before class time.

c) If a monkey sees an act, then that monkey will imitate that act.

e) If a body has infection, then the white corpuscle count increases.

g) If you are at least 16 years of age and pass a driver's test, then you can legally drive a car.

3.

p	q	r	r	$r \wedge \bar{r}$	$\bar{q}$	$p \wedge \bar{q}$	M $(p \wedge \bar{q}) \rightarrow (r \wedge \bar{r})$	$p \rightarrow q$	M $\rightarrow$ $(p \rightarrow q)$
T	T	T	F	F	F	F	T	T	T
T	T	F	T	F	F	F	T	T	T
T	F	T	F	F	T	T	F	F	T
T	F	F	T	F	T	T	F	F	T
F	T	T	F	F	F	F	T	T	T
F	T	F	T	F	F	F	T	T	T
F	F	T	F	F	T	F	T	T	T
F	F	F	T	F	T	F	T	T	T

4. a) T

b) F

c) F

d) F

e) T

f) F

g) T

h) T

i) T

j) T

5. a)

p	q	$\bar{p}$	$\bar{q}$	$\bar{p} \to \bar{q}$
T	T	F	F	T
T	F	F	T	T
F	T	T	F	F
F	F	T	T	T

d)

p	q	$q \to p$	$\overline{q \to p}$
T	T	T	F
T	F	T	F
F	T	F	T
F	F	T	F

h)

p	q	$\bar{p}$	$p \to \bar{p}$	$q \to (p \to \bar{p})$
T	T	F	F	F
T	F	F	F	T
F	T	T	T	T
F	F	T	T	T

l)

p	q	$\bar{p}$	$\bar{q}$	$\bar{q} \to \bar{p}$	$p \to q$	$(\bar{q} \to \bar{p}) \wedge (p \to q)$
T	T	F	F	T	T	T
T	F	F	T	F	F	F
F	T	T	F	T	T	T
F	F	T	T	T	T	T

n)

p	q	r	$\bar{r}$	$\bar{r} \to p$	M $q \wedge (\bar{r} \to p)$	$p \vee r$	$\overline{p \vee r}$	$\overline{p \vee r} \to M$
T	T	T	F	T	T	T	F	T
T	T	F	T	T	T	T	F	T
T	F	T	F	T	F	T	F	T
T	F	F	T	T	F	T	F	T
F	T	T	F	T	T	T	F	T
F	T	F	T	F	F	F	T	F
F	F	T	F	T	F	T	F	T
F	F	F	T	F	F	F	T	F

p)

p	q	r	$q \wedge p$	$q \vee p$	M $(q \wedge p) \rightarrow (q \vee p)$	$r \rightarrow q$	$M \wedge (r \rightarrow q)$
T	T	T	T	T	T	T	T
T	T	F	T	T	T	T	T
T	F	T	F	T	T	F	F
T	F	F	F	T	T	T	T
F	T	T	F	T	T	T	T
F	T	F	F	T	T	T	T
F	F	T	F	F	T	F	F
F	F	F	F	F	T	T	T

7. a) $p \wedge (r \rightarrow q)$
b) $(q \rightarrow p) \wedge (\bar{q} \rightarrow \bar{p})$
c) $p \rightarrow q$
d) $\bar{p} \vee (p \rightarrow q)$
e) $(\bar{p} \vee q) \rightarrow (r \wedge p)$

Exercises 2.10, 11, 12

1. a) T
b) T
c) T
d) F
e) T
f) T
g) T
h) T
i) F
j) F
k) T
l) T
m) F

3. a) If Shakespeare deserved his fame, then his works should be read by everyone.

b) If Shakespeare's works should be read by everyone, then he deserved his fame.

c) If Shakespeare did not deserve his fame, then his works should not be read by everyone.

d) If Shakespeare's works should not be read by everyone, then he did not deserve his fame.

4. a) If Helen can find her glasses, then she can drive.

b) If Helen can drive, then she can find her glasses.

c) If Helen cannot find her glasses, then she cannot drive.

d) If Helen cannot drive, then she cannot find her glasses.

6. a) The hatchet is dull, and you will not be wasting your efforts chopping the wood.

b) Visitors can make themselves at home and not know where the kitchen is.

c) Carl wants to exemplify good manners, and he should put his elbows on the table or start eating his dessert before the hostess sits down.

d) The sun goes down, and the wind blows, and the people at Candlestick Park neither get extremely cold nor go home.

Exercises 2.13, 14

1. a) $p \wedge \overline{q}$

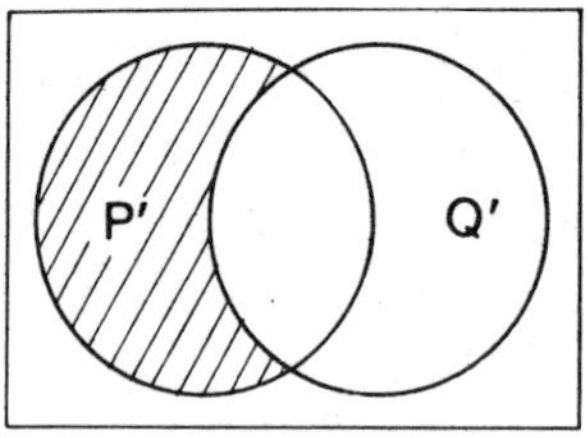

b) $p \wedge (q \wedge \overline{p})$

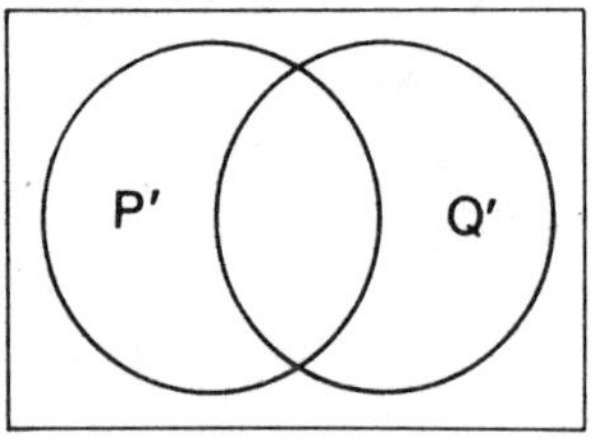

c) $(q \wedge \overline{p}) \Longrightarrow (q \vee p)$

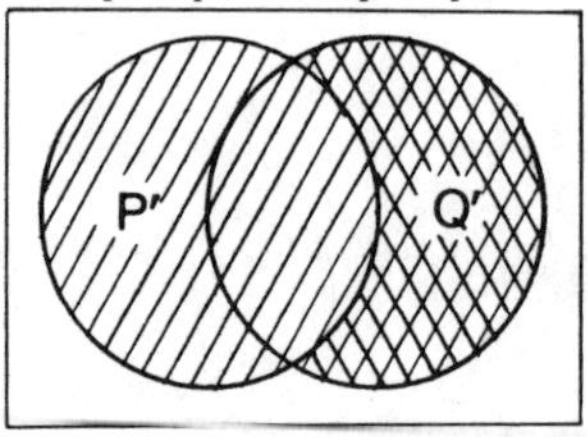

d) $q \wedge \overline{(p \wedge q)} \Longrightarrow q$

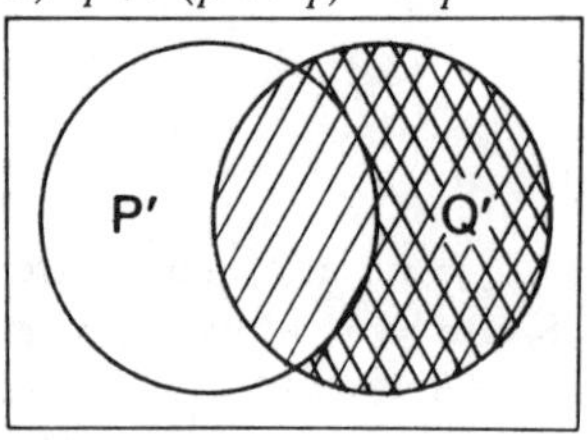

e) $(\overline{p} \wedge \overline{q}) \wedge \overline{r} \Longleftrightarrow \overline{(p \vee r) \vee q}$

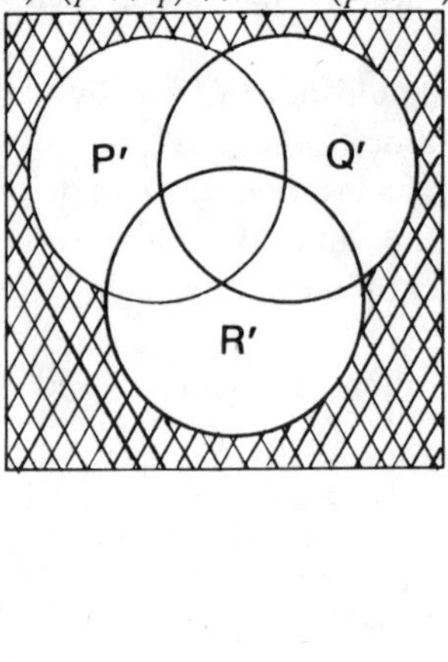

f) $[(r \vee p) \wedge (p \wedge \bar{r})] \Rightarrow [(p \vee r) \wedge (\overline{p \wedge r})]$

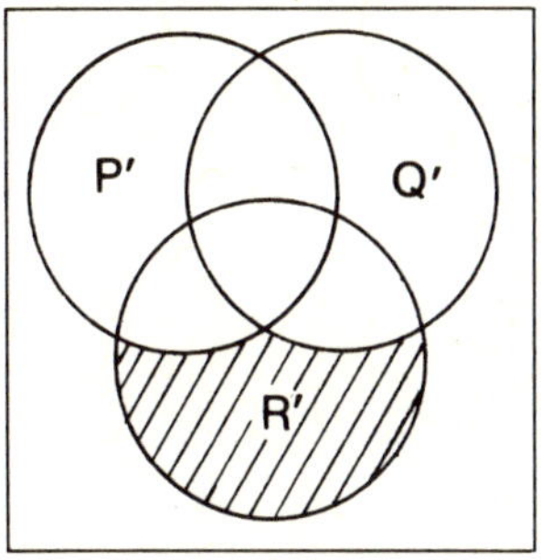

g) $r \wedge [\overline{\bar{q} \wedge (p \wedge \bar{r})}]$

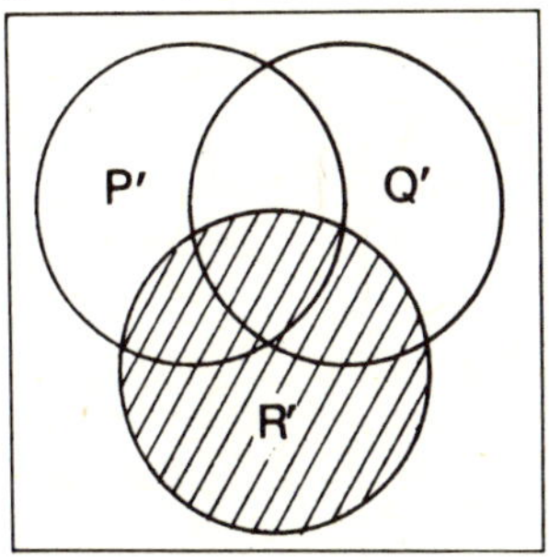

h) $[q \wedge (\bar{p} \vee \bar{r})] \wedge [\overline{p \wedge (q \wedge p)}]$

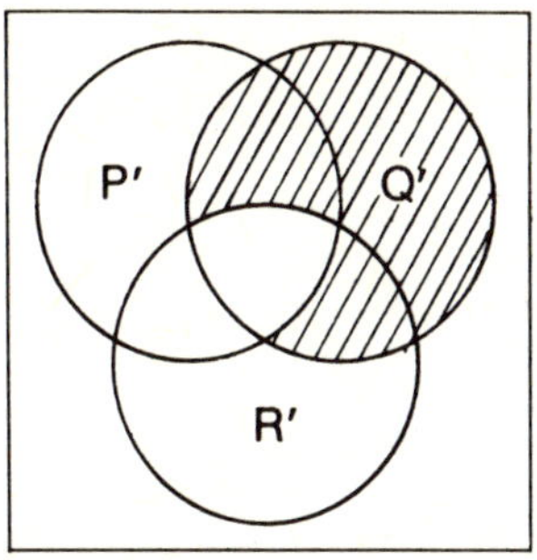

i) $[(p \vee r) \wedge \bar{p}] \wedge \bar{q}$

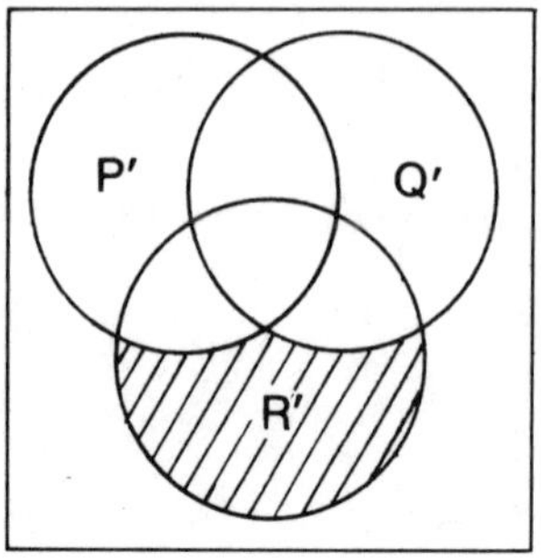

Exercises 3.1

1. a) $p = \overline{P}, q = \overline{Q}$ Law of detachment
b) $p = P \wedge Q \wedge S, q = R \wedge T$ Law of detachment
c) $p = P, q = \overline{Q}$ Absorption
d) $p = P, q = \overline{Q}$ Addition
e) $p = P, q = Q, r = R, s = S$ Constructive dilemma
f) $p = P \wedge Q, r = S \wedge T, s = U, q = R$ Destructive dilemma
g) $p = P \vee Q, q = S \vee T$ Alternative argument
h) $p = P, q = Q \rightarrow (S \vee T)$ Law of detachment
i) $p = P, q = Q \wedge R$ Indirect argument
j) $p = P \wedge Q, q = R$ Conjunction

Exercises 3.2, 3, 4

1. a) $[(P \rightarrow Q) \wedge (Q \rightarrow R) \wedge (R \rightarrow S)] \Rightarrow (P \rightarrow S)$

Statements	Reasons
1. $P \rightarrow Q$	1. Premise
2. $Q \rightarrow R$	2. Premise
3. $P \rightarrow R$	3. 1,2 and Law of syllogism
4. $R \rightarrow S$	4. Premise
5. $P \rightarrow S$	5. 3,4 and Law of syllogism

c) $[(P \rightarrow Q) \wedge (Q \rightarrow R) \wedge P] \Rightarrow R$

Statements	Reasons
1. $P \rightarrow Q$	1. Premise
2. $Q \rightarrow R$	2. Premise
3. $P \rightarrow R$	3. 1,2 and Law of syllogism
4. P	4. Premise
5. R	5. 3,4 and Law of detachment

e) $[(P \rightarrow Q) \wedge (Q \rightarrow R) \wedge \overline{Q} \wedge \overline{R}] \Rightarrow (\overline{P} \wedge \overline{Q})$

Statements	Reasons
1. $P \rightarrow Q$	1. Premise
2. $Q \rightarrow R$	2. Premise
3. $P \rightarrow R$	3. 1,2 and Law of syllogism
4. $\overline{R}$	4. Premise
5. $\overline{P}$	5. 3,4 and Indirect argument
6. $\overline{Q}$	6. Premise
7. $\overline{P} \wedge \overline{Q}$	7. 5,6 and Conjunction

g) $[(P \vee Q) \wedge (R \rightarrow \overline{P}) \wedge R] \Rightarrow Q$

Statements	Reasons
1. $R \rightarrow \overline{P}$	1. Premise
2. R	2. Premise
3. $\overline{P}$	3. 1,2 and Law of detachment
4. $P \vee Q$	4. Premise
5. Q	5. 3,4 and Alternative argument

i) $[P \wedge (P \rightarrow Q) \wedge R] \Rightarrow (P \wedge Q \wedge R)$

Statements	Reasons
1. $P \rightarrow Q$	1. Premise
2. P	2. Premise
3. Q	3. 1,2 and Law of detachment
4. R	4. Premise
5. $P \wedge Q \wedge R$	5. 2,3,4 and Conjunction

k) $[(P \to Q) \wedge \bar{P}] \Rightarrow \bar{Q}$

p	q	$\bar{p}$	$\bar{q}$	$p \to q$	$(p \to q) \wedge \bar{p}$	$[(p \to q) \wedge \bar{p}] \to \bar{q}$
T	T	F	F	T	F	T
T	F	F	T	F	F	T
F	T	T	F	T	T	F
F	F	T	T	T	T	T

2. a) $P \to Q$
$\bar{Q}$
$\therefore \bar{P}$

Statements	Reasons
1. $P \to Q$	1. Premise
2. $\bar{Q}$	2. Premise
3. $\bar{P}$	3. 1,2 and Indirect argument

c) $P \to (Q \wedge S)$
$Q \to R$
$S \to T$
P
$\therefore R \wedge T$

Statements	Reasons
1. $P \to (Q \wedge S)$	1. Premise
2. P	2. Premise
3. $Q \wedge S$	3. 1, 2 and Law of detachment
4. $Q \to R$	4. Premise
5. Q	5. 3 and Simplification
6. R	6. 4, 5 and Law of detachment
7. $S \to T$	7. Premise
8. S	8. 3 and Simplification
9. T	9. 7, 8 and Law of detachment
10. $R \wedge T$	10. 6, 9 and Conjunction

e) $Q \rightarrow P$
$R \rightarrow S$
$\overline{P} \vee \overline{S}$
$\therefore \overline{Q} \vee \overline{R}$

Statements	Reasons
1. $Q \rightarrow P$	1. Premise
2. $R \rightarrow S$	2. Premise
3. $\overline{P} \vee \overline{S}$	3. Premise
4. $\overline{Q} \vee \overline{R}$	4. 1, 2, 3, and Destructive dilemma

Exercises 3.5

1. $[(P \vee \overline{Q}) \wedge Q] \Rightarrow P$

Statements	Reasons
1. $(P \vee \overline{Q}) \wedge Q$	1. Premise
2. $P \vee \overline{Q}$	2. 1 and Simplification
3. Q	3. 1 and Simplification
4. P	4. 2, 3 and Alternative argument

3. $[(P \rightarrow Q) \wedge (Q \rightarrow R)] \Rightarrow (\overline{R} \rightarrow \overline{P})$

Statements	Reasons
1. $(P \rightarrow Q) \wedge (Q \rightarrow R)$	1. Premise
2. $P \rightarrow Q$	2. 1 and Simplification
3. $Q \rightarrow R$	3. 1 and Simplification
4. $P \rightarrow R$	4. 2, 3 and Law of syllogism
5. $\overline{R} \rightarrow \overline{P}$	5. 4 and Contrapositive

5. $[P \rightarrow (Q \wedge R)] \Rightarrow [(P \rightarrow Q) \wedge (P \rightarrow R)]$

Statements	Reasons
1. $P \rightarrow (Q \wedge R)$	1. Premise
2. $\overline{P} \vee (Q \wedge R)$	2. 1 and Conditional
3. $(\overline{P} \vee Q) \wedge (\overline{P} \vee R)$	3. 2 and Distributive
4. $(P \rightarrow Q) \wedge (P \rightarrow R)$	4. 3 and Conditional

7. $[(P \wedge \overline{Q}) \rightarrow \overline{P}] \Rightarrow (P \rightarrow Q)$

Statements	Reasons
1. $(P \wedge \overline{Q}) \rightarrow \overline{P}$	1. Premise
2. $\overline{(P \wedge \overline{Q})} \vee \overline{P}$	2. 1 and Conditional
3. $(\overline{P} \vee Q) \vee \overline{P}$	3. 2, De Morgan's Law and Double negation
4. $\overline{P} \vee Q$	4. 3 and Reduction
5. $P \rightarrow Q$	5. 4 and Conditional

9. $\overline{P} \Rightarrow \overline{(P \wedge Q)}$

Statements	Reasons
1 $\overline{P}$	1. Premise
2. $\overline{P} \vee \overline{Q}$	2. 1 and Addition
3. $\overline{P \wedge Q}$	3. 2 and De Morgan's Law

11. $[(P \rightarrow Q) \wedge (R \rightarrow Q)] \Rightarrow [(P \vee R) \rightarrow Q]$

Statements	Reasons
1. $(P \rightarrow Q) \wedge (R \rightarrow Q)$	1. Premise
2. $(\overline{P} \vee Q) \wedge (\overline{R} \vee Q)$	2. 1 and Conditional
3. $(\overline{P} \wedge \overline{R}) \vee Q$	3. 2 and Distributive
4. $(\overline{P \vee R}) \vee Q$	4. 3 and De Morgan's Law
5. $(P \vee R) \rightarrow Q$	5. 4 and Conditional

13. $\{[P \rightarrow (Q \rightarrow R)] \wedge (\overline{S} \vee P) \wedge Q\} \Rightarrow (S \rightarrow R)$

Statements	Reasons
1. $[P \rightarrow (Q \rightarrow R)] \wedge (\overline{S} \vee P) \wedge Q$	1. Premise
2. $\overline{S} \vee P$	2. 1 and Simplification
3. $S \rightarrow P$	3. 2 and Conditional
4. $P \rightarrow (Q \rightarrow R)$	4. 1 and Simplification
5. $S \rightarrow (Q \rightarrow R)$	5. 3, 4 and Law of syllogism
6. $\overline{S} \vee (\overline{Q} \vee R)$	6. 5 and Conditional

continued

Statements	Reasons
7. $(\bar{S} \vee R) \vee \bar{Q}$	7. 6 and Associativity and Commutativity
8. Q	8. 1 and Simplification
9. $\bar{S} \vee R$	9. 7, 8 and Alternative Argument
10. $S \rightarrow R$	10. 9 and Conditional

15. $\{[(P \wedge Q) \rightarrow S] \wedge [(Q \rightarrow S) \rightarrow (R \vee T)] \wedge (\bar{R} \wedge \bar{T})\} \Rightarrow \bar{P}$

Statements	Reasons
1. $[(P \wedge Q) \rightarrow S] \wedge [(Q \rightarrow S) \rightarrow (R \vee T)] \wedge (\bar{R} \wedge \bar{T})$	1. Premise
2. $(Q \rightarrow S) \rightarrow (R \vee T)$	2. 1 and Simplification
3. $\overline{(R \vee T)} \rightarrow \overline{(Q \rightarrow S)}$	3. 2 and Contrapositive
4. $(\bar{R} \wedge \bar{T}) \rightarrow \overline{(Q \rightarrow S)}$	4. 3 and De Morgan's Law
5. $\bar{R} \wedge \bar{T}$	5. 1 and Simplification
6. $\overline{Q \rightarrow S}$	6. 4, 5 and Law of detachment
7. $Q \wedge \bar{S}$	7. 6, Conditional and De Morgan's Law
8. $\bar{S}$	8. 7 and Simplification
9. $(P \wedge Q) \rightarrow S$	9. 1 and Simplification
10. $\overline{P \wedge Q}$	10. 8, 9 and Indirect argument
11. $\bar{P} \vee \bar{Q}$	11. 10 and De Morgan's Law
12. Q	12. 7 and Simplification
13. $\bar{\bar{Q}}$	13. 12 and Double Negation
14. $\bar{P}$	14. 11, 13 & Alternative argument

Exercises 3.6

1. $\overline{S}$

$\therefore \overline{S \wedge P}$

Statements	Reasons
1. $\overline{S}$	1. Premise
2. $\overline{S} \vee \overline{P}$	2. 1 and Addition
3. $\overline{S \wedge P}$	3. 2 and De Morgan's Law

3. $R \to \overline{P}$
$M \to R$
$\overline{T} \to P$

$\therefore M \to T$

Statements	Reasons
1. $R \to \overline{P}$	1. Premise
2. $\overline{T} \to P$	2. Premise
3. $\overline{P} \to T$	3. 2 and Contrapositive
4. $R \to T$	4. 1, 3 and Law of syllogism
5. $M \to R$	5. Premise
6. $M \to T$	6. 5, 4 and Law of syllogism

5. $L \vee \overline{C}$
$(\overline{A} \vee \overline{D}) \to \overline{L}$

$\therefore C \to (A \wedge D)$

Statements	Reasons
1. $L \vee \overline{C}$	1. Premise
2. $\overline{L} \to \overline{C}$	2. 1 and Conditional
3. $(\overline{A} \vee \overline{D}) \to \overline{L}$	3. Premise
4. $(\overline{A} \vee \overline{D}) \to \overline{C}$	4. 2, 3 and Law of syllogism
5. $(\overline{A \wedge D}) \to \overline{C}$	5. 4 and De Morgan's Law
6. $C \to (A \wedge D)$	6. 5 and Contrapositive

7. $(F \vee C) \rightarrow (D \wedge O)$
$(O \vee R) \rightarrow V$
$\overline{}$
$\therefore F \rightarrow V$

Statements	Reasons
1. $(O \vee R) \rightarrow V$	1. Premise
2. $(\overline{O \vee R}) \vee V$	2. 1 and Conditional
3. $(\bar{O} \wedge \bar{R}) \vee V$	3. 2 and De Morgan's Law
4. $(\bar{O} \vee V) \wedge (\bar{R} \vee V)$	4. 3 and Distributive
5. $(O \rightarrow V) \wedge (R \rightarrow V)$	5. 4 and Conditional
6. $O \rightarrow V$	6. 5 and Simplification
7. $(F \vee C) \rightarrow (D \wedge O)$	7. Premise
8. $[(F \vee C) \rightarrow D] \wedge$ $[(F \vee C) \rightarrow O]$	8. 7, #5, page 168
9. $(F \vee C) \rightarrow O$	9. 8 and Simplification
10. $(\overline{F \vee C}) \vee O$	10. 9 and Conditional
11. $(\bar{F} \wedge \bar{C}) \vee O$	11. 10 and De Morgan's Law
12. $(\bar{F} \vee O) \wedge (\bar{C} \vee O)$	12. 11 and Distributive
13. $(F \rightarrow O) \wedge (C \rightarrow O)$	13. 12 and Conditional
14. $F \rightarrow O$	14. 13 and Simplification
15. $F \rightarrow V$	15. 6, 14 and Law of syllogism

9. $(K \rightarrow W) \wedge (T \rightarrow H)$
$(W \rightarrow C) \wedge (H \rightarrow M)$
$\overline{C \wedge M}$
$K \rightarrow T$
$\overline{}$
$\therefore \bar{K}$

Statements	Reasons
1. $(K \rightarrow W) \wedge (T \rightarrow H)$	1. Premise
2. $(W \rightarrow C) \wedge (H \rightarrow M)$	2. Premise
3. $K \rightarrow W$	3. 1 and Simplification
4. $W \rightarrow C$	4. 2 and Simplification
5. $K \rightarrow C$	5. 3, 4 & Law of syllogism
6. $T \rightarrow H$	6. 1 and Simplification
7. $H \rightarrow M$	7. 2 and Simplification

continued

Statements	Reasons
8. $T \to M$	8. 6, 7 & Law of syllogism
9. $\overline{C \wedge M}$	9. Premise
10. $\overline{C} \vee \overline{M}$	10. 9 and De Morgan's Law
11. $\overline{K} \vee \overline{T}$	11. 5, 8, 10 & Destructive dilemma
12. $K \to T$	12. Premise
13. $\overline{T} \to \overline{K}$	13. 12 and Contrapositive
14. $K \to \overline{T}$	14. 11 and Conditional
15. $K \to \overline{K}$	15. 13, 14 & Law of syllogism
16. $\overline{K} \vee \overline{K}$	16. 15 and Conditional
17. $\overline{K}$	17. 16 and Reduction

Exercises 4.1

1. a) "$y = x + 1$"
c) "y is the first two letters of the name of the day x."
e) "y is the color of the fruit x."
g) "y is an antonym of x."

3. a) {(1,5), (2,6), (3,7), (4,8), (5,9), (6,10)}
c) {(2,5), (4,8)}
e) {(1,2), (3,3), (5,4), (7,5), (9,6)}

4. a) No
b) Yes
c) No
d) Yes
e) No
f) Yes
g) No
h) No

5. a) $\mathfrak{D}$ = {red, yellow, green} $\mathfrak{R}$ = {apple, banana, lime}
c) $\mathfrak{D}$ = {0, 1, 2, 3} $\mathfrak{R}$ = {8, 11, 14, 17}

6. a, b, c, f, h, j, k, l

Exercises 4.2

1. R^{-1} = {(1,x), (2,y), (3,z)}

2. {(1,1)}, Yes

3. (1) {(a,a), (b,b), (c,c), (d,d)}; Yes
(2) {(a,a), (b,b), (c,c), (d,d), (a,b), (b,a)}; Yes
(3) {(a,a), (b,b), (c,c), (d,d), (a,c), (c,a)}; Yes
(4) {(a,a), (b,b), (c,c), (d,d), (b,c), (c,b)}; Yes

(5) {(a,a), (b,b), (c,c), (d,d), (c,d), (d,c)}; Yes

4. Yes

5. a) $a = b + 2$ e) $a = \frac{12 - b}{5}$

c) $a = \sqrt{b}$

Exercises 4.3

1. a) $\mathfrak{D} = \{3,4,5,6\}$ $\mathcal{R} = \{-1,12,-2\}$

c) $\mathfrak{D}$ = {happy, joy, glad} $\mathcal{R}$ = {glad, delighted}

2. c, d, e

3. a) 49

c) 2, −7

e) 83

5. Show: $f(A \cap B) \subset f(A) \cap f(B)$

Let $y \in f(A \cap B)$. This means $y = f(c)$ where $c \in A \cap B$. Thus, $c \in A$ and $c \in B$. From this we get $f(c) \in f(A)$ and $f(c) \in f(B)$ so that $f(c) \in f(A) \cap f(B)$. Hence, $y \in f(A) \cap f(B)$ and thus, $f(A \cap B) \subset f(A) \cap f(B)$.

Exercises 4.4

1. a, b, f, h, i

2. a, b, f, h

3. a)

a	$f(a)$
1	3
2	8
3	15
4	24
5	35
6	48
7	63
8	80

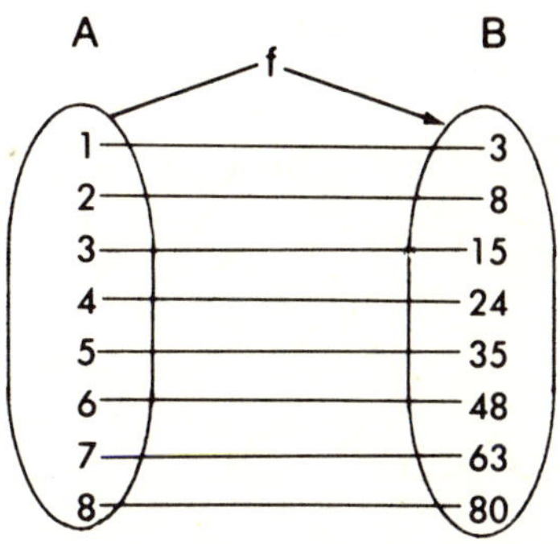

c)

a	*b*
12	7
4	8
3	6
−1	7
9	12
13	4

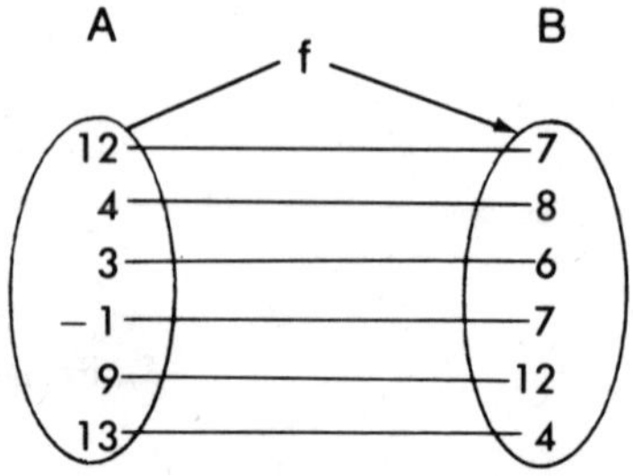

4. b)

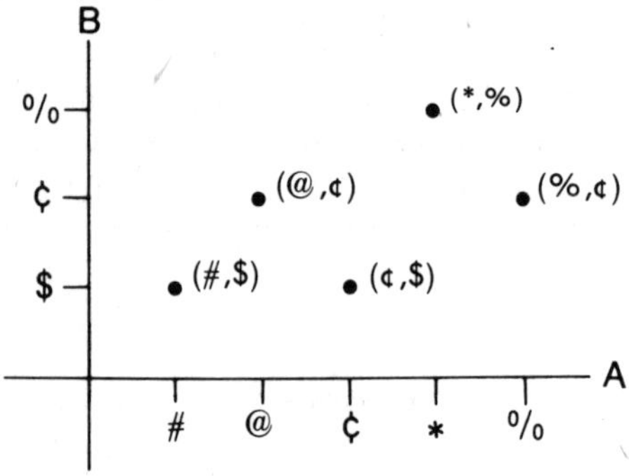

Exercises 4.5

1. a, d, e, f, k

   2. b) $\{(x,y) \mid y = x + 2 \text{ where } x, y \in \mathbf{R}\}$

x	y
−2	0
−1	1
0	2
1	3

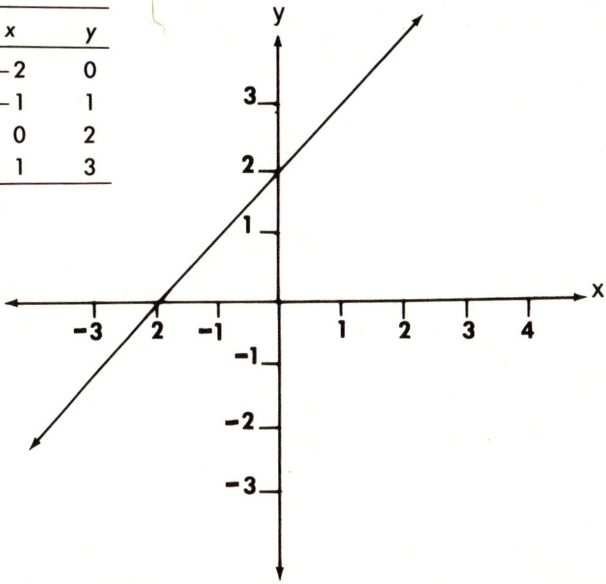

d) $\{(x,y) \mid y = 3x - 4 \text{ where } x,y \in \mathbf{R}\}$

x	y
0	−4
1	−1
2	2
3	5

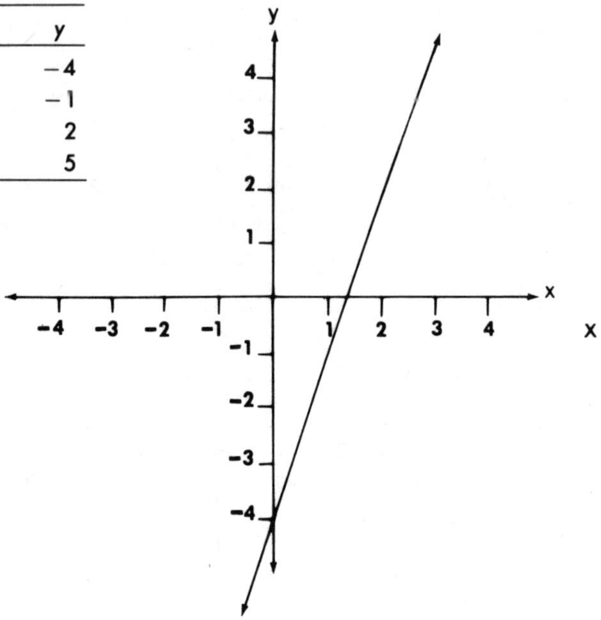

3. a) I b) II c) III

4.

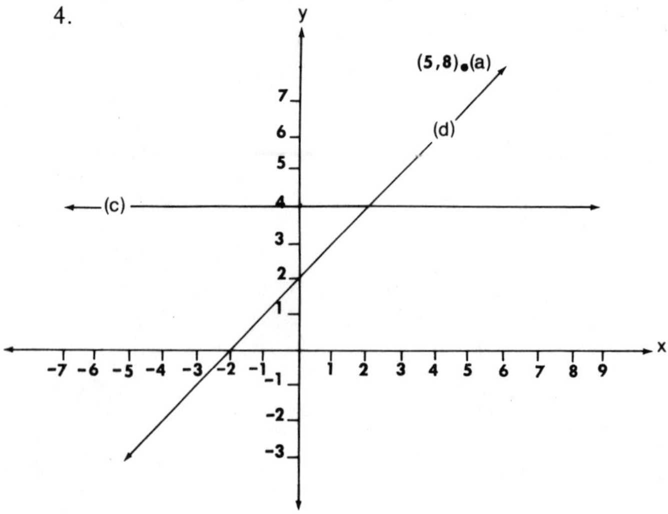

Exercises 4.6

2. a) $f^{-1} = \{(5,3), (6,1), (-2,5), (1,6)\}$ Yes
c) $f^{-1} = \{(y,x) \mid x = \sqrt{y + 9} \text{ where } x \in Z\}$ No
e) $f^{-1} = \{(b,a) \mid b \in \{2,4,6,8,10\} \text{ and } b = a\}$ Yes

Exercises 4.7

1. a) Yes c) Yes e) Yes
b) No d) Yes f) No
2. 6
3. 4 from {a, b} into itself; 2 of which are onto, 2 are 1–1.
4. Yes
5. Yes, No
6. No
7. Yes
8. a) Let $f(n) = f(m)$ which implies $5n = 5m$ which implies $n = m$

b) Let $q \in B$, then $q = 5n$ and $n = \frac{q}{5}$ for $n \in A$, i.e., for each element of B there exists an element of A which corresponds to it.

c) $f^{-1}: B \rightarrow A$ by the rule: If $q \in B$ then $f^{-1}(q) = \frac{q}{5}$

9. a) $f: A \rightarrow B$ by the rule: If $n \in A$ then $f(n) = n + 2$
c) $f: A \rightarrow B$ by the rule: If $n \in A$ then $f(n) = 2n + 2$

Exercises 4.8

1. a)

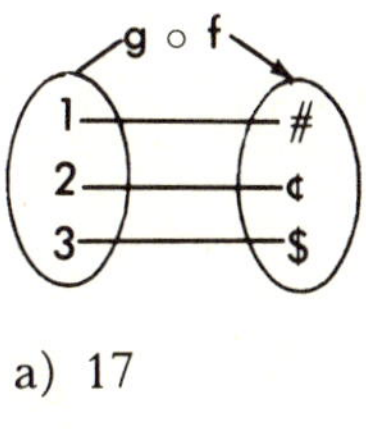

b) $\{(1,\#), (2,¢), (3,\$)\}$
No

2. a) 17 e) 26
d) 39 h) 10
4. b) $(g \circ f)(x) = 15x + 18$
c) $f \circ f(x) = 9x + 16$

5. a) $f \circ g(x) = x^4 - 2x^2 - 1$
d) $g \circ g(x) = x^4 - 2x^2$
8. a) α
b) α
c) Associativity

Exercises 4.9

1. $f \circ f^{-1}(1) = f(f^{-1}(1)) = f(a) = 1$
$f \circ f^{-1}(2) = f(f^{-1}(2)) = f(b) = 2$
$f \circ f^{-1}(3) = f(f^{-1}(3)) = f(c) = 3$

Exercises 4.10

1. a) Not commutative, Not associative
c) Commutative, Associative
e) Commutative, Not associative
2. a) Not; Consider $(2,4) \in Q \times Q$. Then $2 * 4 = \sqrt{2+4} = \sqrt{6}$ But $\sqrt{6} \notin Q$.
b) Not; Consider $(3,2) \in Z \times Z$. Then $3 * 2 = \frac{3}{2}$. But, $\frac{3}{2} \notin Z$
c) Not; Consider $(3,5) \in N \times N$. Then $3 * 5 = 3 - 5 = -2$. But $-2 \notin N$.
d) Is a binary operation.

Exercises 4.11

1. 2, 3, 5, 7, 11, 13, 17, 19, 23, 29, 31, 37, 41, 43, 47
2. a) 6 c) 0 e) 0
3. a) 10 c) 1 e) 8
4. b, c, d, e, f, g
6.

$\oplus$	0
0	0

7. Yes, Yes

8.

⟐	1	2	3	4
1	1	2	3	4
2	2	4	1	3
3	3	1	4	2
4	4	3	2	1

9.

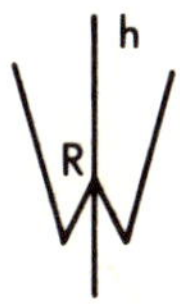

*	I	H
I	I	H
H	H	I

H = Flip about line h.
I = No change.

11. a) $\begin{pmatrix} 1 & 2 & 3 & 4 & 5 \\ 4 & 1 & 2 & 5 & 3 \end{pmatrix}$ c) $\begin{pmatrix} 1 & 2 & 3 & 4 & 5 \\ 5 & 2 & 4 & 3 & 1 \end{pmatrix}$

12. a) $\begin{pmatrix} 1 & 2 & 3 & 4 & 5 \\ 4 & 2 & 1 & 5 & 3 \end{pmatrix}$ c) $\begin{pmatrix} 1 & 2 & 3 & 4 & 5 \\ 3 & 1 & 5 & 4 & 2 \end{pmatrix}$

13.

$$f_1 = \begin{pmatrix} A & B & C \\ A & C & B \end{pmatrix} \qquad f_2 = \begin{pmatrix} A & B & C \\ C & B & A \end{pmatrix}$$

$$f_3 = \begin{pmatrix} A & B & C \\ B & A & C \end{pmatrix} \qquad f_4 = \begin{pmatrix} A & B & C \\ B & C & A \end{pmatrix}$$

$$f_5 = \begin{pmatrix} A & B & C \\ C & A & B \end{pmatrix} \qquad f_6 = \begin{pmatrix} A & B & C \\ A & B & C \end{pmatrix}$$

$\circ$	f_1	f_2	f_3	f_4	f_5	f_6
f_1	f_6	f_5	f_4	f_3	f_2	f_1
f_2	f_4	f_6	f_5	f_1	f_3	f_2
f_3	f_5	f_4	f_6	f_2	f_1	f_3
f_4	f_2	f_3	f_1	f_5	f_6	f_4
f_5	f_3	f_1	f_2	f_6	f_4	f_5
f_6	f_1	f_2	f_3	f_4	f_5	f_6

14. f_1 = flip about h
f_2 = flip about k
f_3 = flip about l
f_4 = flip about m
f_5 = flip about n
f_6 = rotation about 0 of 72°
f_7 = rotation about 0 of 144°
f_8 = rotation about 0 of 216°
f_9 = rotation about 0 of 288°
f_{10} = rotation about 0 of 360°

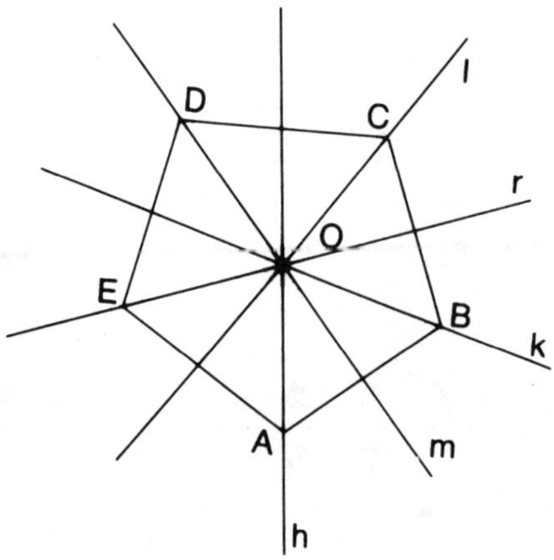

$$f_1 = \begin{pmatrix} A & B & C & D & E \\ A & E & D & C & B \end{pmatrix} \qquad f_2 = \begin{pmatrix} A & B & C & D & E \\ C & B & A & E & D \end{pmatrix}$$

$$f_3 = \begin{pmatrix} A & B & C & D & E \\ E & D & C & B & A \end{pmatrix} \qquad f_4 = \begin{pmatrix} A & B & C & D & E \\ B & A & E & D & C \end{pmatrix}$$

$$f_5 = \begin{pmatrix} A & B & C & D & E \\ D & C & B & A & E \end{pmatrix} \qquad f_6 = \begin{pmatrix} A & B & C & D & E \\ B & C & D & E & A \end{pmatrix}$$

$$f_7 = \begin{pmatrix} A & B & C & D & E \\ C & D & E & A & B \end{pmatrix} \qquad f_8 = \begin{pmatrix} A & B & C & D & E \\ D & E & A & B & C \end{pmatrix}$$

$$f_9 = \begin{pmatrix} A & B & C & D & E \\ E & A & B & C & D \end{pmatrix} \qquad f_{10} = \begin{pmatrix} A & B & C & D & E \\ A & B & C & D & E \end{pmatrix}$$

15. f_1 = flip about h
f_2 = flip about j
f_3 = flip about l
f_4 = flip about k
f_5 = rotation of 90°
f_6 = rotation of 180°
f_7 = rotation of 270°
f_8 = rotation of 360°

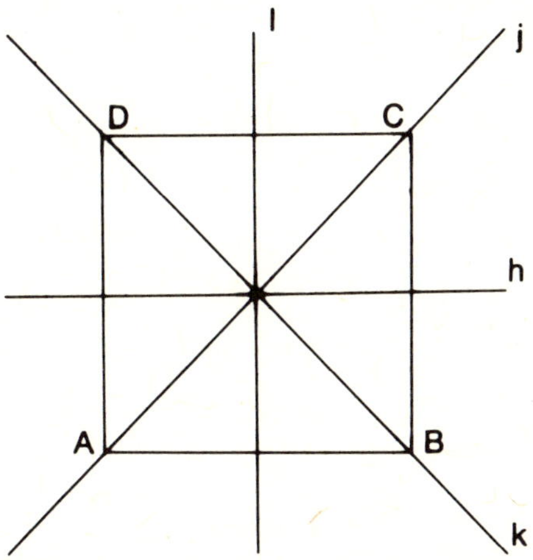

$\circ$	f_1	f_2	f_3	f_4	f_5	f_6	f_7	f_8
f_1	f_8	f_5	f_6	f_7	f_2	f_3	f_4	f_1
f_2	f_7	f_8	f_5	f_6	f_3	f_4	f_1	f_2
f_3	f_6	f_7	f_8	f_5	f_4	f_1	f_2	f_3
f_4	f_5	f_6	f_7	f_8	f_1	f_2	f_3	f_4
f_5	f_4	f_1	f_2	f_3	f_6	f_7	f_8	f_5
f_6	f_3	f_4	f_1	f_2	f_7	f_8	f_5	f_6
f_7	f_2	f_3	f_4	f_1	f_8	f_5	f_6	f_7
f_8	f_1	f_2	f_3	f_4	f_5	f_6	f_7	f_8

Exercises 5.1

1. Yes
2. Yes
3. a, c, e, f, i, k
4. No
5. No
6. Yes
7. a) No, no identity
   b) Yes
   c) No, no identity
   d) Yes
9. Yes
10. Yes, Yes
   a) $I \Diamond I = I$, $R \Diamond R = I$, $H \Diamond H = I$, $V \Diamond V = I$
   b) $I \Diamond I \Diamond I = I$
   c) $I \Diamond R = R$, $R \Diamond I = R$, $H \Diamond V = R$, $V \Diamond H = R$
11. No

Exercises 5.2

1. Premises: $(G, *)$ is a group and $b * a = b * c$ for a, b, c elements of G.

   Conclusion: $a = c$

Statements	Reasons
1. $b * a = b * c$	1. Premise
2. $b \in G$ means $b' \in G$	2. G_3 (Inverse $*$)
3. $b' * (b * a) = b' * (b * c)$	3. Definition of $=$
4. $b' * (b * a) = (b' * b) * a$ $b' * (b * c) = (b' * b) * c$	4. G_1 (Associative $*$)
5. $b' * (b * a) = e * a$ $b' * (b * c) = e * c$	5. G_3 (Inverse $*$)
6. $b' * (b * a) = a$ $b' * (b * c) = c$	6. G_2 (Identity $*$)
7. $a = c$	7. 3,6 and Definition of $=$

2. Premises: $(G, *)$ is a group and there are at least two distinct inverses, a_1 and a_2, relative to $*$.

Conclusion: Contradiction

Statements	Reasons
1. a_1 and a_2 are distinct inverses relative to $*$; i.e., $a_1 \neq a_2$	1. Premise
2. For $a \in G$, $a * a_1 = e$ and $a * a_2 = e$	2. G_3 (Inverse $*$)
3. $a * a_1 = a * a_2$	3. Definition of $=$
4. $a_1 = a_2$	4. Theorem 1
5. $a_1 \neq a_2$ and $a_1 = a_2$	5. 1 and 4
6. Contradiction	6. Principle of logic. Statement of form $r \wedge \bar{r}$ on step 5.

3. Using a' with the binary operation $*$ on the left of both sides of $y * a = b$ yields $y = b * a'$, which is a solution since $y * a = (b * a') * a = b * (a' * a) = b * e = b$. To show that there is not more than one solution to $y * a = b$, we assume that there exists y_1 and y_2 such that both are solutions to $y * a = b$. That is, we assume $y_1 * a = b$ and $y_2 * a = b$. This means that $y_1 * a = y_2 * a$. But then, by the right cancellation law $y_1 = y_2$, so that there is exactly one solution to $y * a = b$.

4. a) Yes; a = 2, b = 3
b) Yes; a = 4, b = 2
c) No

6. a) Yes
b) Yes
c) Let a = f_2 and b = f_5, then $(f_2 \circ f_5)' = f_4' = f_4$, and $f_2' \circ f_5' = f_3 \circ f_5 = f_6$. Hence, $(f_2 \circ f_5)' \neq f_2' \circ f_5'$. But, $(f_2 \circ f_5)' = f_4' = f_4$ and $f_5' \circ f_2' = f_5 \circ f_3 = f_4$. Hence, $(f_2 \circ f_5)' = f_5' \circ f_2'$

7. Premises: $(G, *)$ a group, $a, b \in$ G and $a = b$
Conclusion: $a' = b'$

Statements	Reasons
1. $a,b \in$ G means $a',b' \in$ G	1. Definition of a group
2. $a = b$	2. Premise
3. $a * a' = b * a'$	3. Definition of =
4. $e = b * a'$	4. G_3 (Inverse $*$)
5. $b' * e = b' * (b * a')$	5. Definition of =
6. $b' * e = (b' * b) * a'$	6. G_1 (Associative $*$)
7. $b' * e = e * a'$	7. G_3 (Inverse $*$)
8. $b' = a'$	8. G_2 (Identity $*$)

8. a' with the binary operation $*$ on the left of both sides of $a * x * b = c$ and b' with the binary operation $*$ on the right of both sides of $a * x * b = c$ yields $x = (a' * c) * b'$, which is a solution since $a * x * b =$ a $* [(a' * c) * b'] * b = [(a * a') * c] * (b' * b) = (e * c) * e = c * e = c$. To show that there is not more than one solution to $a * x * b = c$, we assume that there exists x_1 and x_2 such that both are solutions of $a * x * b = c$. That is, we assume $a * x_1 * b = c$ and $a * x_2 * b = c$. This means $a * x_1 * b = a * x_2 * b$. But then, by the left and right cancellation laws, $x_1 = x_2$ so that there is exactly one solution to $a * x * b = c$.

Exercise 5.3

1. a) $(\{0,2\}, \oplus)$
b) $(\{I,R\},\diamond)$, $(\{I,V\},\diamond)$, $(\{I,H\},\diamond)$

c) $(\{f_1, f_4\}, \circ)$, $(\{f_1, f_6\}, \circ)$; $(\{f_1, f_5\}, \circ)$, $(\{f_1, f_2, f_3\}, \circ)$
d) None
e) $(\{1,6\}, \diamond)$

3. $\{e\}$

4. a) $(\{f_5, f_6, f_7, f_8\}, \circ)$, $(\{f_6, f_8\}, \circ)$, $(\{f_4, f_8\}, \circ)$, $(\{f_2, f_8\}, \circ)$, $(\{f_1, f_8\}, \circ)$, $(\{f_8\}, \circ)$, $(\{f_3, f_8\}, \circ)$, $(\{f_1, f_3, f_6, f_8\}, \circ)$, $(\{f_2, f_4, f_6, f_8\}, \circ)$, $(\{f_1, f_2, f_3, f_4, f_5, f_6, f_7, f_8\}, \circ)$ where f_i is as defined in the answer to Problem 15, Exercise 4.11.
b) $(\{f_5, f_6, f_7, f_8\}, \circ)$ and $(\{f_2, f_4, f_6, f_8\}, \circ)$

5. Since $(H, *)$ and $(K, *)$ are subgroups of the group $(G, *)$, then the identity e of $(G, *)$ is in $(H, *)$ and in $(K, *)$. That is, $e \in H$ and $e \in K$. Thus, $e \in H \cap K$ and hence $H \cap K$ is not empty. Now, let a and b (not necessarily distinct) be elements of $H \cap K$. Then, $a, b \in H$ and $a, b \in K$. Since $a, b \in H$ then $a * b' \in H$. Similarly, $a * b' \in K$. Thus, $a * b' \in H \cap K$ and hence $(H \cap K, *)$ is a subgroup of $(G, *)$.

6. Consider the subgroups $(\{I, R\}, \diamond)$ and $(\{I, H\}, \diamond)$ of the group $(S, \diamond)$. (See Figure 5.8) Now, $(\{I, R\} \cup \{I, H\}, \diamond)$ is the same as $(\{I, R, H\}, \diamond)$, which is not a subgroup of $(S, \diamond)$.

8. Since $e * e = e^2 = e$ for $e \in G$, we know that $(H, *)$ is not empty. Now suppose $a, b \in H$. Then $a^2 = e$ and $b^2 = e$. So $a^2 = b^2$ or $a * a = b * b$. Now,

$$\begin{aligned}
(a * a) * b' &= (b * b) * b' \\
(b' * a) * (a * b') &= (b' * b) * (b * b') \\
(a * b') * (a * b') &= e * e \\
(a * b')^2 &= e
\end{aligned}$$

Hence, $(a * b') \in H$ and thus $(H, *)$ is a subgroup of $(G, *)$.

9. a) $(S, \diamond)$
b) $(\{1,4\}, \diamond)$
c) $(\{0\}, +)$
d) $(\{1, -1\}, \cdot)$

Exercises 5.4

1. a) 2
c) 3
e) infinite
g) 2
i) infinite

2. a) 2

c) 3

e) infinite

g) 2

i) infinite

3. a) 0

b) 1 or 5

c) 2 or 4

d) 3

4. $(T,\circ)$

5. a) 1, 2, 3, 4

c) 1, 3, 5, 7, 9, 11, 13, 15

7. 3, −3

8. 1, 2, 4, 8, 16

Exercises 5.5

1. a)

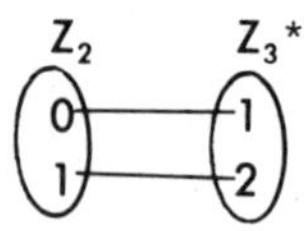

c) 0—rotation of 360°
1—rotation of 270°
2—rotation of 180°
3—rotation of 90°

d) A—rotation of 360°
B—rotation of 180°
C—flip about h
D—flip about l
E—flip about k
F—flip about j
(C–F: As in the answer to Problem 15, Exercise 4.11)
G—rotation of 90°
H—rotation of 270°

2. No

4. Since $(G,*)$ and $(G',*)$ are isomorphic, there exists a function $f:G \to G'$ such that f is one-to-one and onto. Hence, $n(G) = n(G')$.

7. a) $(Z_5^*, \odot)$

b) With rigid motions of a square, f_8, f_2, f_4, f_6 as in the answer to Problem 15, Exercise 4.11, $(\{f_8, f_2, f_4, f_6\}, \circ)$.

c) No

9. Must show: i) function, ii) one-to-one, iii) onto, and iv) preserves operation

i) Suppose $a = b$, then $a' = b'$, and since $b' = f(b)$ and $a' = f(a)$ this means $f(a) = f(b)$ so that f is a function.

ii) Let $f(a)$ and $f(b)$ be elements of G such that $f(a) = f(b)$. Then, $a' = b'$ and $a'' = b''$ so that $a = b$ and f is one-to-one.

iii) Let $a' \in G$. Since $(G, *)$ is a group, $\exists a \in G$ such that $a * a' = e$. Hence $\exists a \in G$ so that $f(a) = a'$ and f is onto.

iv) Let $a, b \in G$. Then, $f(a * b) = (a * b)' = b' * a' = a' * b' = f(a) * f(b)$ since $(G, *)$ is Abelian. Hence, f preserves the operation $*$.

Exercises 5.6

1. $a * b = a$. Since $(G, *)$ is a group, $\exists e \in G$ such that $a * e = a$ for $a \in G$ and e is unique. Now, $a * b = a * e$ and, by the left cancellation law, $b = e$.

2. Let $a, b \in H$. Then $a', b' \in H$ and since $*$ is a binary operation, $a * b' \in H$. Hence, $(H, *)$ is a subgroup of $(G, *)$.

3. Let $a, b \in G$. Now, $a * a = e$ and $b * b = e$ and $(a * b) * (a * b) = e$ so that $a = a'$, $b = b'$ and $(a * b) = (a * b)'$. Hence, $a * b = b' * a' = b * a$ and so $(G, *)$ is Abelian.

6. $(Q^+, \cdot)$, $(\{1\}, \cdot)$, $\left(\left\{\frac{1}{3n} \mid n \in Z\right\}, \cdot\right)$, $\left(\left\{\frac{1}{2n} \mid n \in Z\right\}, \cdot\right)$

7. Let $a, b \in G$. Show $a * b = b * a$. $f(a * b) = f(a) * f(b) = a' * b' = (b * a)' = f(b * a)$. Since f is one-to-one, $a * b = b * a$, and hence, $(G, *)$ is Abelian.

Exercises 6.1

2. No.

3.

+	0	1	2	3	4
0	0	1	2	3	4
1	1	2	3	4	0
2	2	3	4	0	1
3	3	4	0	1	2
4	4	0	1	2	3

·	0	1	2	3	4
0	0	0	0	0	0
1	0	1	2	3	4
2	0	2	4	1	3
3	0	3	1	4	2
4	0	4	3	2	1

5. No, the distributive properties do not hold.
6. No, the left distributive law does not hold.
7. Yes, $1 = b$

·	a	b	c
a	a	a	a
b	a	b	c
c	a	c	b

8. Yes
9. $2(1 + 3) = 2 \cdot 0 = 0 \neq 2 = 0 + 2 = 2 \cdot 1 + 2 \cdot 3$
10. Yes, with the universal set being the identity.
11. No, does not satisfy both left and right distributive laws.
12. No

Exercises 6.2

3. $(a + b) + [-(a + b)] = 0$ since $-(a + b)$ is the inverse of $(a + b)$.
But,

$$\begin{aligned}(a + b) + [-a - b] &= (a + b) + [-a + (-b)] \\ &= a + [b + (-a)] + (-b) \\ &= a + [(-a) + b] + (-b) \\ &= [a + (-a)] + [b + (-b)] \\ &= 0 + 0 \\ &= 0\end{aligned}$$

Hence, $-a - b$ is also the inverse of $(a + b)$. So $-(a + b) = -a - b$.

4.

$$\begin{aligned} a(b - c) &= a[b + (-c)] \\ &= ab + a(-c) \\ &= ab + (-ac) \\ &= ab - ac \end{aligned}$$

6. a) $a = a^2 = (a)(a) = (-a)(-a) = -a$

6. b)

$$\begin{aligned} a + b &= (a + b)^2 \\ a + b &= a^2 + ab + ba + b^2 \\ a + b &= a + ab + ba + b \\ 0 &= ab + ba \end{aligned}$$

But then, $-ab = ba$, and since $-ab = ab$ by Part 1, so $ab = ba$.

7. $a \cdot a^{-1} = 1$ where a^{-1} is the inverse of a. Suppose a' is also an inverse and $a' \neq a^{-1}$ then $a' \cdot a = a \cdot a' = 1$ for any $a \in$ R. So, $a^{-1} \cdot a' = 1$ and $a' \cdot a^{-1} = 1$ and,

$$\begin{aligned} a^{-1} \cdot a &= a' \cdot a \\ (a^{-1} \cdot a) \cdot a' &= (a' \cdot a) \cdot a' \\ a^{-1} \cdot (a \cdot a') &= a' \cdot (a \cdot a') \\ a^{-1} \cdot 1 &= a' \cdot 1 \\ a^{-1} &= a' \end{aligned}$$

9. $a \cdot a^{-1} = 1$ and $b \cdot b^{-1} = 1$. Now, $(ab)(ab)^{-1} = 1$ and

$$\begin{aligned} (ab)(b^{-1}a^{-1}) &= a(bb^{-1})a^{-1} \\ &= a(1)a^{-1} \\ &= a \cdot a^{-1} \\ &= 1 \end{aligned}$$

So both $(ab)^{-1}$ and $(b^{-1}a^{-1})$ are inverses of (ab) and hence, $(ab)^{-1} = b^{-1}a^{-1}$ since the inverse is unique.

10. Suppose $0 = 1$. Now, since R has at least two distinct elements, there is an $a \in$ R such that $a \neq 0$. So,

$$a \cdot 0 = a \cdot 1$$
$$0 = a$$

But, $a \neq 0$. Hence, a contradiction. Thus, $0 \neq 1$.

Exercises 6.3

1. $(2Z, +, \cdot)$ and $(3Z, +, \cdot)$

2. $(\{0,4,8\}, \oplus, \odot)$, $(\{0,6\}, \oplus, \odot)$, $(\{0,3,6,9\}, \oplus, \odot)$

3. Since $0 \in R'$ and $0 \in R''$, then $0 \in R' \cap R''$, so $R' \cap R''$ is not empty. Now, let $a, b \in R' \cap R''$. This means $a, b \in R'$ and $a, b \in R''$. So $a \in R'$ and $b \in R'$ means $a - b \in R'$ and $ab \in R'$. Similarly, $a \in R''$ and $b \in R''$ means $a - b \in R''$ and $ab \in R''$. Thus, $a - b \in R' \cap R''$ and $ab \in R' \cap R''$. So, by Theorem 4, $(R' \cap R'', +, \cdot)$ is a subring of $(R, +, \cdot)$

4. Since $a \cdot 0 = 0$ for all $a \in R$, $0 \in R_a$, so R_a is not empty. Now, let $x, y \in R_a$. Then $ax = 0$ and $ay = 0$. But, $a(x - y) = ax - ay = 0 - 0 = 0$, so $x - y \in R_a$. Also, $a(xy) = (ax)y = 0 \cdot y = 0$, so $xy \in R_a$. Hence, $(R_a, +, \cdot)$ is a subring of $(R, +, \cdot)$.

5. $(\{0,6\}, \oplus, \odot)$

Exercises 6.4

1. Since $ba = ca$, we know $ba - ca = ca - ca = 0$. But, $ba - ca = (b - c)a$. So that $(b - c)a = 0$. Since $a \neq 0$ and $(R, +, \cdot)$ is an integral domain, we have $b - c = 0$. So $b = c$.

2. a) $\{2,3,4\}$
b) None
c) None
d) $\{2,3,4,6,8,9,10\}$

3. $(R, +, \cdot)$ has at least two elements since $0 \in R$ and $1 \in R$. Now, we must show that, if $ab = 0$, then $a = 0$ or $b = 0$. Suppose $a \neq 0$. Then, since $a \cdot 0 = 0$, we can write $a \cdot 0 = ab$. By the cancellation law, $0 = b$, so $(R, +, \cdot)$ is an integral domain.

4. R′ has at least two elements, 0 and 1. Now, we must prove that, if $ab = 0$, then $a = 0$ or $b = 0$ where $a, b \in R'$. Now, $a \in R'$ means $a \in R$ and $b \in R'$ means $b \in R$. But since $(R, +, \cdot)$ is an integral domain, $ab = 0$ means $a = 0$ or $b = 0$. So $(R', +, \cdot)$ is a commutative ring with identity and has no zero divisors. Hence, $(R', +, \cdot)$ is an integral domain.

Exercises 6.5

1. a,b

3. Since $(R, +, \cdot)$ is a field, it is an integral domain. Hence, by Problem 4, Exercise 6.4, $(R', +, \cdot)$ is an integral domain.

6. By the definition of set G, if $u \in G$, then $u^{-1} \in G$. Since 1 is its own multiplicative inverse in R, then $1 \in G$. If $x, y \in G$, then $x^{-1}, y^{-1} \in R$. But, $(R, +, \cdot)$ a ring means that $y^{-1}x^{-1} \in R$ and $(xy)y^{-1}x^{-1} \in R$. But, $y^{-1}x^{-1} = (xy)^{-1}$. Hence, xy has an inverse in R so that $xy \in G$. Associativity is inherited in $(G, \cdot)$ since $(R, +, \cdot)$ is a ring.

7. a) For $(Z, +, \cdot)$ we would have $G = \{1, -1\}$. Now, $(\{1, -1\}, \cdot)$ is a group.
b) For $(Z_{12}, \oplus, \odot)$, we would have $G = \{1, 5, 7, 11\}$. Now, $(\{1, 5, 7, 11\}, \odot)$ is a group.

Exercises 6.6

4. Hint: Note that $f(a) = f(a \cdot 1)$ for $a \in R$.

5. Must prove that $f(R) \neq \phi$ and, if $a, b \in f(R)$, then $a - b$ and ab are elements of $f(R)$. Note that for $0 \in R$, $f(0) \in f(R)$, so $f(R) \neq \phi$. For $a, b \in f(R)$ there exists $x, y \in R$ such that $f(x) = a$ and $f(y) = b$. Hence,

$$f(x - y) = f(x) - f(y) = a - b \in f(R)$$
$$f(x \cdot y) = f(x) \cdot f(y) = a \cdot b \in f(R)$$

so that $f(R)$ is a subring of S.

6. Hint: Note that $f(1) = f(a \cdot a^{-1})$ for $a \in R$.

Index